This book is to be returned
the last date

Applied Sensory Analysis of Foods

Volume I

Editor

Howard Moskowitz, Ph.D.
Principal
Moskowitz/Jacobs, Inc.
Valhalla, New York

CRC Press, Inc.
Boca Raton, Florida

Library of Congress Cataloging-in-Publication Data

Applied sensory analysis of foods.

Includes bibliography and index.
1. Food — Sensory evaluation. I. Moskowitz,
Howard R.
TX546.A67   1988   664'.07   87-25015
ISBN 0-8493-6705-0 (v. 1)
ISBN 0-8493-6706-9 (v. 2)

D
664.07

MOS

Direct all inquiries to CRC Press, Inc., 2000 Corporate Blvd., N.W., Boca Raton, Florida, 33431.

© 1988 by CRC Press, Inc.

International Standard Book Number 0-8493-6705-0 (v. 1)
International Standard Book Number 0-8493-6706-9 (v. 2)

Library of Congress Card Number 87-25015
Printed in the United States

PREFACE

We witness today an exuberant outpouring of research and practical methods of sensory analysis. Growth, however, did not progress in a well-ordered, linear fashion. Beginning in the early 1900s as an informal assessment of food quality, sensory analysis matured in a desultory manner. Articles on procedure appeared in one or another journal, and the development of "methodology" (the science of method) proceeded at a similar langorous pace. By 1965, however, the pace began to quicken, signalled by the publication of *Principles of Sensory Evaluation of Food* (Amerine, M. A., Pangborn, R. M., and Roessler, E. B., Academic Press, New York, 1965.), the first major book in the newly burgeoning field.

Today, two decades later, the sensory analyst working in both basic research and in a corporate environment confronts an embarrassment of riches. Methods for testing food abound, tangible evidence of the application of published academic research from diverse disciplines such as food science, psychophysics (experimental psychology), and statistics.

These two volumes of *Applied Sensory Analysis* present the field of sensory analysis in its full vigor. They give the reader first-hand descriptions of research approaches and applications, written by experts in the field. These volumes provide a wealth of viewpoints in sensory analysis, in a depth otherwise unobtainable by the reader. Each paper provides a detailed analysis of approaches, and relevant examples, often of practical business issues, addressed and solved by the approach.

The two volumes divide naturally into three categories. "Descriptive Analysis and the Dimensions of Sensory Perception" comprises the first category and contains five papers. These papers lie at the foundation of sensory analysis, for they discuss alternative methods for developing and using language to describe the sensory nuances of foods. The second category, "Difference Testing and Intensity Scaling", presents six papers dealing with basic measurement tools. These papers cover alternative methods, including new ones such as signal detection theory, which derive from other disciplines and only recently have entered the domain of sensory analysis. The third category, "Multivariate Methods", comprises four chapters and deals with new approaches, such as multidimensional scaling to represent stimuli in a geometric space, and the experimental design and optimization of products and concepts.

These volumes thus present a spectrum of approaches, designed both to introduce the novice to the wealth of approaches, and to invite the professional to try these new techniques.

Howard R. Moskowitz

THE EDITOR

Howard R. Moskowitz, Ph.D., is a principal with the marketing research company, Moskowitz/Jacobs, Inc., Valhalla, New York, and the Taste and Smell Consulting Group, Inc., Evanston, Illinois. He received the Ph.D. degree (1969) in Experimental Psychology from Harvard University. Dr. Moskowitz is a member of numerous organizations, including the Institute of Food Technologists, American Society for Testing and Materials, and American Chemical Society, and serves on the editorial board of various journals. In 1972 he founded *Chemical Senses*, an international journal devoted to smell and taste research. In addition, he has published extensively in both scientific and business journals, and is a frequent speaker at many technical and product research meetings.

CONTRIBUTORS

Volume I

Dov Basker, C.Chem., F.R.S.C.
Senior Scientist
Department of Food Science
Agricultural Research Organization
The Volcani Center
Bet Dagan, Israel

Virginia B. Ferguson, M.S.
Director of Panel Training
Food Industries Section
Arthur D. Little, Inc.
Cambridge, Massachusetts

Barry E. Jacobs, M.A.
Principal
Moskowitz/Jacobs, Inc.
Valhalla, New York

David A. Kendall, B.A.
Senior Research Consultant
Food Industries Section
Arthur D. Little, Inc.
Cambridge, Massachusetts

Alan S. Kornheiser, Ph.D.
Senior Vice President/Research Director
The Saugatuck Group
Westport, Connecticut

Ronald S. Leight, Ph.D.
Project Scientist
Department of Physical and Sensory
  Science
International Flavors and Fragrances
Union Beach, New Jersey

Morten C. Meilgaard, M.S., D.Sc.
Vice President of Research
The Stroh Brewery Company
Detroit, Michigan

Howard R. Moskowitz, Ph.D.
Principal
Moskowitz/Jacobs, Inc.
Valhalla, New York

Anne J. Neilson, A.B.
Senior Consultant
Food Industries Section
Arthur D. Little, Inc.
Cambridge, Massachusetts

Michael O'Mahony, Ph.D., M.R.S.C.
Associate Professor
Department of Food Science and
  Technology
University of California
Davis, California

Jacqueline H. Pearce, M.B.A., R.D.
Group Manager
Department of Sensory Evaluation
Quaker Oats Company
Barrington, Illinois

Dwight R. Riskey, Ph.D.
Vice President, Marketing Research and
  New Products
Department of Marketing
Frito-Lay
Plano, Texas

Elaine Z. Skinner, B.S.
Senior Laboratory Manager
Department of Product Evaluation
General Foods Corporation
White Plains, New York

Hal Sokolow, M.A.
President
Applied Behavioral Dynamics
Lake Peekskill, New York

Craig B. Warren, Ph.D.
Vice President and Director
Organoleptic Research Department
International Flavors and Fragrances
Union Beach, New Jersey

Katherine L. Zook, M.Sc.
Consultant
Lake Barrington Shores
Barrington, Illinois

TABLE OF CONTENTS

Volume I

DESCRIPTIVE ANALYSIS AND THE DIMENSIONS OF SENSORY PERCEPTION

Chapter 1
Qualitative Methods for Language Development 3
Hal Sokolow

Chapter 2
Profile Methods: Flavor Profile and Profile Attribute Analysis 21
Anne J. Neilson, Virginia B. Ferguson, and David A. Kendall

Chapter 3
Quantitative Descriptive Analysis 43
Katherine L. Zook and Jacqueline H. Pearce

Chapter 4
Beer Flavor Terminology — A Case History... 73
Morten C. Meilgaard

Chapter 5
The Texture Profile Method 89
Elaine Z. Skinner

DIFFERENCE TESTING AND INTENSITY SCALING

Chapter 6
Difference Testing: Procedures and Panelists.. 111
Alan S. Kornheiser

Chapter 7
Assessor Selection: Procedures and Results 125
Dov Basker

Chapter 8
Sensory Difference and Preference Testing: The Use of Signal
Detection Measures.. 145
Michael O'Mahony

Chapter 9
Uses and Abuses of Category Scales in Sensory Measurement 177
Dwight R. Riskey

Chapter 10
Magnitude Estimation: Scientific Background and Use in Sensory Analysis............ 193
Howard R. Moskowitz and Barry E. Jacobs

Chapter 11
Standing Panels Using Magnitude Estimation for Research and
Product Development...225
Ronald S. Leight and Craig B. Warren

Index 253

TABLE OF CONTENTS

Volume II

MULTIVARIATE METHODS

Chapter 12
Basic Concepts of Multidimensional Scaling 3
Susan S. Schiffman

Chapter 13
Computer-Aided Sensory Evaluation of Foods and Beverages.......................... 35
Armand V. Cardello, Owen Maller, and Mary V. Klicka

Chapter 14
Experimental Design and Analysis 83
Maximo C. Gacula, Jr.

Chapter 15
Simultaneous Optimization of Products and Concepts for Foods 141
Howard R. Moskowitz and Barry E. Jacobs

Index 175

Descriptive Analysis and the Dimensions of Sensory Perception

Chapter 1

QUALITATIVE METHODS FOR LANGUAGE DEVELOPMENT

Hal Sokolow

TABLE OF CONTENTS

I.   Introduction 4

II.   Qualitative Research Designs 5
       A.   Focus Groups 5
       B.   In-Depth Interviews 5
       C.   Mini-Groups 6
       D.   Maxi-Groups 6
       E.   Dual-Moderated Groups 6
       F.   Respondent-Plus-Client Groups 6

III.   Moderating and Group Dynamics 6
       A.   Group Dynamics 6
       B.   Characteristics of the Moderator 7

IV.   Techniques for Enhancing Group Productivity 8
       A.   Task Orientation 8
       B.   Open-Ended Questioning 8
       C.   Structuring Questions for Simplicity 8
       D.   Committed Responses in Writing 8
       E.   Stimulus Dimension Presensitization 9
       F.   Specific Probes for Richer Responses 9
       G.   Relaxation Training 10
       H.   Free Associaton 10
       I.   Product Imagery and Projective Techniques 10
       J.   Nonverbal Language 11
       K.   Presession Orientation and Diaries 11
       L.   Show and Tell 11
       M.   Preinterview Home Usage 11

V.   Selecting Respondents 12
       A.   Product Usage 12
       B.   Nonusers 12
       C.   Psychographic/Lifestyle Groups 12
       D.   Verbal Fluency 12
       E.   Creative Ability 12
       F.   Screening Out Professional Respondents 12

VI.   Considerations for a Systematic Study Design 13
       A.   Standardization Across Groups 13
       B.   Controlling the Environment 13
       C.   Timing the Research 13
       D.   Frames of Reference: Monadic and Comparative Evaluations 13

E.   Cleansing the Palate.. 13
F.   The Limits of Multiple Tastings....................................... 14

VII.   Framework for Organizing a Consumer Language Study 14
A.   The Combination Study: Language and Product Dynamics.............. 14
B.   The Pure Consumer Language Study 15
C.   Language Development Procedure...................................... 15
D.   Language Development Instructions 16
1.   Monadic Instructions.. 16
2.   Comparative Instructions.. 16
3.   Language Development Form...................................... 16

VIII.   Applying Language Findings to Quantitative Testing 17

I. INTRODUCTION

Practical product development is ultimately founded upon the satisfaction of consumer needs. Proper understanding of these needs is dependent upon clear communication with consumers through a common language. The communication challenge begins with the fact that the everyday language of the consumer and that of the food scientist are quite different. People generally tend to use a limited, inconsistent, and relatively nonspecific vocabulary to describe their perceptions. Food scientists, on the other hand, require comprehensive, consistent, and very specific terminology in order to fine-tune product dimensions into new product designs.

Unless a definite research approach and questioning sequence is thoroughly planned in advance of a product study, the resulting information will be confusing and of minimal value. Consumer's confusion with questions and difficulty with tasks must also be anticipated. Consumers need to be guided in an organized manner throughout the study so that they can understand what is required of them. Ambiguities in questioning and in responses are a constant concern in interviewing consumers and can make the difference between a productive study and one that is meaningless.

Consumer discussions need to be focused. When asked for taste reactions to a soft drink, a respondent is apt to use generalizations such as "I like it/I don't like it", "It's good/It's bad", or "It's strong/It's flat/It's heavy". Although the extent of liking is an important part of any evaluation, the specifics of *why* the person likes or doesn't like the product are crucial information for technical guidance. Vague taste descriptors such as "strong", "flat", or "heavy" are meaningless unless qualified by additional references or replaced by more accurate terms. For example, does "strong" for a beverage refer to too much flavor or too much aftertaste? Furthermore, what *kind* of flavor or aftertaste is being referred to? Does "flat" mean not enough carbonation or not enough flavor? Does "heavy" mean too much syrup or too cloudy? Often, the meanings of such words go beyond two or three possible interpretations.

Fortunately, these apparent difficulties can be overcome, and standardized consumer research techniques can be applied throughout the entire research and development process. Consumer language research can enhance product development by:

1. Providing a better understanding of what attributes mean in technical terms
2. Building more consistent attribute lists for product evaluation
3. Initiating a systematic approach for improving consumer acceptance of products

II. QUALITATIVE RESEARCH DESIGNS

Qualitative research is an exploratory tool that provides the flexibility and open-ended response opportunities that are particularly helpful for language development. Qualitative research usually takes the form of focus groups, in-depth interviews, or a combination of both. In addition, there are other, less frequently used designs which can be used for special applications. These include mini-groups, maxi-groups, dual-moderated groups, and respondent-plus-client groups.

A. Focus Groups

A typical focus group consists of ten people who have been prerecruited to fit specific demographic, attitudinal, and usage characteristics. In return for a monetary incentive (cooperation fee), these respondents assemble at a central facility conference room which has a one-way mirror. On the other side of the mirror is a viewing room that allows interested project personnel to observe the group dynamics and feedback without being seen. The focus group discussion is led by a professional moderator who is experienced in group interviewing techniques and has sufficient knowledge of the category under discussion.

The moderator follows a preplanned discussion guide to ensure full coverage of all issues and product dimensions. Since the main format of a focus group session is open-ended, the moderator can probe as needed to follow up and clarify any comments from consumers. If necessary, observers in the viewing room can communicate with the moderator by passing message notes to the moderator or by an electronic telecommunication setup using an earphone for the moderator. Such communication may be desirable when the client wishes to elaborate on the current discussion or explore new topics and hypotheses inspired by the immediately observable feedback.

Although qualitative research generally offers the useful advantages of flexibility in approach and unlimited probing, the choice of focus groups, in-depth interviews, or other qualitative approach can be based on the consideration of additional benefits and special situations.

Focus groups offer several advantages over in-depth interviews. In focus groups, ideas and language generation tend to be stimulated by the interchange of comments. By reacting in agreement and disagreement to each other's responses, group members find new frames of reference for assessing their own responses. This stimulation from the group often helps respondents to express their own feelings. Focus groups are also less costly on a per-respondent basis because they require less time to execute and analyze. These advantages tend to make focus groups the most popular choice for language generation and product development research.

B. In-Depth Interviews

Logistically, in-depth interviews are conducted in a similar way to focus groups, except that the interviews are individual (one-to-one) rather than in a group setting. The main advantages of the in-depth approach are the elimination of peer group influence bias and the generation of more personal and honest responses. By removing group pressure, respondents are more likely to reveal their own feelings, rather than those considered most acceptable among peers. The language obtained from each respondent is therefore a true reflection of his own perceptions rather than a parroting of others' responses in a group situation.

This advantage of the in-depth approach becomes particularly relevant in exploring attitudes and usage in categories where consumers are inclined to provide socially acceptable responses rather than those reflective of their actual behavior. This applies in the area of alcoholic beverages, for example, where real and reported experiences may be inconsistent. Dieting and health consciousness are two similar areas where people's words and actions are often disparate.

Another type of situation where individual responses may be useful is in understanding true feelings among consumer targets that are particularly sensitive to group influence. Teenagers, particularly boys, tend to be overly self-conscious in groups, and are often reluctant to express their own feelings in front of others. Yea-saying to majority opinion and agreement with more outspoken group members tend to be more frequent with this age group. Individual interviews can eliminate these distorting influences. The disadvantages of in-depths are that they require more research and analytical time, as well as their incremental cost.

C. Mini-Groups

Mini-groups consist of four to five respondents and can be an effective compromise between full focus groups and in-depth interviews. Mini-groups provide a practical solution when a large number of groups need to be researched to isolate diversified screening requirements. Compared to full focus groups, mini-groups encourage more balanced participation among respondents, yea-saying is lessened, and individual patterns of behavior become more apparent.

D. Maxi-Groups

Maxi-groups consist of 15 to 30 respondents and are appropriate for more structured questioning sequences where immediate follow-up and flexibility are still important. A series of such groups can be combined to provide qualitative flexibility with an opportunity for statistical analysis.

E. Dual-Moderated Groups

Dual-moderated groups are focus groups using a professional moderator plus a client representative as discussion leaders in the same session. The professional moderator handles guidance of the general group dynamics and basic questioning. The client representative, who may be an expert from research and development, probes technical issues which may be beyond the scope of the moderator's background. Dual-moderated groups are an effective method when technical explanations are necessary before consumers can provide informed responses.

F. Respondent-Plus-Client Groups

Respondent-plus-client groups involve consumers and client respresentatives in separate group sessions. Issues are first discussed with consumers to gain insights and background information such as general attitudes and vocabulary. Thereafter, the moderator leads a discussion with a group of client representatives to review the material and insights just obtained from consumers. At this point, the most useful ideas and findings are identified and are used as the framework for the moderator's subsequent detailed analysis and recommendations.

III. MODERATING AND GROUP DYNAMICS

A. Group Dynamics

The productivity of a focus group session depends on the willingness of respondents to cooperate with the moderator, as well as their ability to make discriminations and express their perceptions. The moderator is responsible for keeping respondents positively motivated throughout the course of each session. The objective is to create an atmosphere where group members feel free and encouraged to contribute their thoughts and feelings for richer feedback.

A cooperative spirit can be established at the beginning of a session by building a comfortable, friendly rapport and conveying an attitude of task importance. Generally, a brief explanation of the reason for the group meeting is adequate. This shows respondents that there is a meaningful purpose to the procedure and that their responses are truly important. The moderator can explain that, because different people like different things, manufacturers are interested in what consumers think about products. This information, it can be explained, is then used to make better products thay they personally will enjoy using.

To clarify the respondent's role in the session, they can be advised that (1) they should express their honest thoughts and feelings, (2) there are no right or wrong answers, and (3) the first thing that comes to mind is usually the most important.

Easy warm-up questions help to build rapport by giving group members an opportunity to adapt to the interviewing situation. An effective beginning is to start going around the discussion table with a simple and short direct question. In this manner each person understands that he or she is expected to give a response, and the discussion momentum is begun. A good starter question of this type is "What brands of _____ do you use, and how often do you use each one?" Once everyone has participated in this round-robin fashion, it becomes easier for group members to volunteer subsequent responses.

Creating an accepting atmosphere during the entire discussion is important in order to allay fears of criticism or embarrassment. Anxieties, such as the common fear of speaking in front of others, affect people's memory and ability to express themselves clearly. The group's anxiety can be reduced if the moderator projects a friendly but in-control attitude. Group members will feel most comfortable if they are not pressed beyond their ability to respond, are not interrupted while trying to respond, and feel that they are genuinely being listened to rather than being interrogated. The moderator should avoid probing any one respondent excessively. If a respondent cannot answer a question even after it is paraphrased, the moderator can say, "If a thought occurs to you later about that, please let me know or write it down." Otherwise, respondents will be reluctant to volunteer for fear of endless questioning that puts them on the spot. Follow-up questions can also be referred to other group members by turning and saying, "How do you feel about that?", thus relieving pressure from a temporarily inarticulate individual.

Balanced participation from all members should be encouraged from the outset because individual response patterns are established early in the session. If a respondent sees that she can avoid participation by avoiding eye contact or by shrugging her shoulders, she will adopt that tactic for the rest of the session. Such people may need a few more direct, easy-to-answer questions to give them the confidence to speak spontaneously.

B. Characteristics of the Moderator

The appropriateness of particular moderators should be taken into account when choosing who will personally conduct the interviews. Personalities, interviewing styles, and experience vary greatly across different moderators and may be more or less effective for certain situations. Moderators may vary in effectiveness with different consumer targets based on the group's age, sex, or ethnic composition. Moderators also vary in their levels of experience with different product categories, research methodologies, and techniques.

The ideal moderator brings a whole-picture perspective to the project by being experienced in marketing, technical, and psychological dynamics. Such a person helps ensure optimal project design by clearly defining the issues and objectives. Complete coverage of the issues can be ensured by thoughtful planning of the moderator's guide. The skilled moderator encourages richer, more meaningful information by knowing how to stimulate the group dynamic process. Valid interpretation of consumer feedback is assured by objectivity in moderation and reporting. The research is made action-oriented by providing specific research and development implications and recommendations. Finally, the ideal moderator has a highly flexible personality. This is necessary to follow up on all clues, to rephrase unclear questions, to improvise alternative methods for getting at difficult information, and for relating to a broad array of respondent personalities. Flexibility is also required to accommodate last-minute or on-the-spot changes in the questioning guide or research methodology.

IV. TECHNIQUES FOR ENHANCING GROUP PRODUCTIVITY

A. Task Orientation

Consumers are not used to being very specific in describing their taste perceptions. Providing a short introductory orientation and emphasizing that descriptions should be as specific as possible can help reduce the number of generalized responses. It should be explained to respondents that vague terms such as "good taste" or "has an aftertaste" are not precise enough to give guidance for product improvement. The hows, whats, and whys really need to be known for each statement. *How* is the taste good, and *what* kind of aftertaste is it, and *why* is it a good or bad aftertaste? Mentioning these expectations in advance will motivate respondents to apply more effort in considering their own impressions, rather than providing the easiest forms of answers.

B. Open-Ended Questioning

Open-ended questions are usually the most difficult for people to answer. People are not always sure of the relevant context of the question or what the parameters of "acceptable" responses are. For example, we may ask a person, "What is that most important thing for a soft drink to have for you to like it?" Although the person will give us an answer, we don't know if the person is thinking of the beverage as a drink by itself or in conjunction with food. If it is drunk by itself, is that in a relaxed situation, or during or immediately after physical activity (such as sports)? If it is consumed with food, is that with a meal or with snacks, and at what time of day? The context of usage is often not clear but usually directly affects motivations for usage and perceptions of the product's characteristics.

The best way to avoid such ambiguity is to ascertain early in the interview what the key usage dynamics are and to qualify responses by occassion and predominant need. A beverage drunk during during relaxation may be enjoyed mainly for its taste. The same product in a high-activity situation may be desired mainly because it is thirst-quenching and goes down smoothly. Whatever consumer needs are predominant in a given usage situation will influence which characteristics become most salient and noticeable.

C. Structuring Questions for Simplicity

Simplification of sentence structure is the key to clearly communicating all specific questions and tasks. People vary greatly in their ability to comprehend questions and articulate responses. Orienting the interviews for communication to the least fluent in the sample assures comprehension by the entire group.

Generally, the more concisely, plainly worded, and immediately relevantly the question can be phrased, the easier it will be to understand and answer. Sometimes, of course, additional explanations will be necessary to fully clarify a more difficult concept. Here again, the additional explanations or paraphrasing should be as concise and plainly worded as possible. Long-winded explanations have a tendency to confuse more than enlighten.

D. Committed Responses in Writing

A very useful method for increasing the productivity of consumer language focus groups is committing responses in writing. Rather than having respondents immediately verbalize their answers in front of the group, they are asked to write down all their impressions first. The committment in writing minimizes group influence bias since the actual responses are written individually. After consumers have recorded their answers, they can be asked, "What are all the words you wrote down to describe the taste of this product?"

In addition to obtaining more honest responses in this manner, the technique provides a good record of the range of each person's responses to a given question or stimulus. This technique also facilitates responses among more introverted types of respondents. Instead of being confronted with "What do you think?", the respondent is asked, "What is written on your paper?"

Committed responses are best obtained using standardized questionnaire forms that correspond to the flow of the group discussion. The forms should be clearly labeled to indicate which responses are to be recorded and where. This makes it easier for respondents to organize their responses and simpler for the moderator to review the materials. Simply handing out blank sheets of paper for recording will create more work for the analyst after the group work is over. When sample sizes are large enough, responses on questionnaire forms can be tallied, and directional interpretations become possible.

E. Stimulus Dimension Presensitization

Respondents can be sensitized to subtle differences by telling them in advance what kinds of general differences they can expect to detect. This focuses their attention on dimensions of interest to the researcher which require further elaboration and specific descriptions to be useful.

Respondents may, for example, be told that the next two products they will taste vary in texture. By presensitizing respondents to this dimension, they are now obligated to define the specifics of texture differences rather than just saying that the products differ in texture. After all preselected dimensions have been elaborated, respondents can be asked if there are any other dimensions that they noticed that they were not questioned on. This form of insurance gives respondents an opportunity to contribute reactions which the researcher may not have anticipated.

F. Specific Probes for Richer Responses

Respondents will attempt to describe their impressions of various products with varying degrees of ease or difficulty. All opportunities should be taken to encourage the most specific descriptors possible and to elaborate the full range of responses. Frequent probes are usually needed to clarify taste perceptions. For example, if a person says a food is tangy, we don't know if that means unpleasantly sour or pleasantly lemon-like.

Probes can be made less repetitious and monotonous by varying their wording. To achieve the desired clarification and identification of specific taste impressions, probe variations such as the following may be used:

1. "When you say that it tastes ____, is it a certain *kind* of ____?"
2. "What other words would closely describe the same feeling?"
3. "What other food items would have a similar ____?"
4. "What is it about this product that makes you say it's ____?"

Elaborative probes can be used to stimulate the generation of additional descriptors. The wording of these probes can also be varied to reduce monotony. Elaborative probes can be phrased in the following manner:

1. "Does tasting and eating this make you think of any other sensations?"
2. "What other qualities and characteristics come to mind when you taste and eat it?"

G. Relaxation Training

Although we do not want to create an unnatural or totally atypical product evaluation situation, the need for finer discriminations may warrant the use of special procedures. One example is guided imagery for heightened sensory awareness and feedback. Verbal suggestions can be given to respondents that will allow them to become completely relaxed. Additional suggestions in this relaxed state can increase their ability to concentrate on the tasks, free their associative process, and enhance their ability to communicate. Such procedures are frequently used in clinical practice to free self-expression. They can also be effective in a focus group situation to enhance spontaneity.

All suggestions are given in a calm, rhythmic manner, subtly reinforcing that the respondents should allow themselves to become calm and quiet, and as they become more and more relaxed they will become more and more sensitive to the impressions that they receive through the senses from this product. Such suggestions direct and concentrate the respondents' attention and tend to phase out other distractions. The relaxed state of mind lets associations come more easily to the fore. These associations may be more personal, more colorful, and more unique. This is because associations in a relaxed state tend to tap into more distant memory impressions, so a person may describe an acidic, highly carbonated taste by saying, "When I was a kid my father once let me put my tongue across two battery terminals — that's the kind of sharp feeling I got from this drink."

Additional suggestions can be given to respondents regarding their ability to communicate their sensations. By giving positively reinforcing suggestions that they will find it easier and easier to find the words to describe their impressions, the respondents are inspired to greater confidence, and performance anxiety is further reduced.

Although the administration of relaxation imagery requires extra patience, the technique is especially helpful in situations where people are unaccustomed to using a wide range of terms to describe a product.

H. Free Association

In the free association technique, respondents are asked to mention or write down all the words that come to mind when they think of a particular stimulus word. This technique works best after a brief relaxation training exercise and after some warm-up word-association exercises with practice words. The respondents are probed for more associations until a certain number of responses has been achieved.

Free association is a basic technique that is very productive for generating long lists of product attributes as well as the connotations associated with these attributes. For example, a food product may generate different associations, and therefore different perceptions, if it is described as "diet" rather than "sugar-free". "Diet" may have negative associations such as "nobody likes to diet" and "dieting is no fun". "Sugar-free" on the other hand may have more positive associations such as "it's healthier for you because it doesn't have sugar" or "it won't give you a lot of calories". For analytical purposes, free associations can be ranked in terms of first associations and subsequent responses.

I. Product Imagery and Projective Techniques

Although the product development process tends to concentrate on the physical characteristics of product attributes, the psychological interpretations are also important. The psychological associations connected with a product will directly affect how the product will be perceived and whether or not it will be accepted.

Colors of foods, for example, may convey strength of flavor; whether they are natural or artificial; or whether they are more suitable for men, women, or children. Including an assessment of product imagery in the research ensures that the more deeply learned associations are added to the immediate sense impressions for a complete profile. The goal is to have physical product attributes and imagery associations in harmony and reinforcing each other. The imagery of a product reflects certain expectations based on previous experience. If these expectations are not fulfilled, consumer disappointment and product rejection will result.

Product imagery can be elicited through direct questions such as, "In what situations and on what occassions would you use this product?" Various projective techniques can also be employed. For example, respondents can be asked, "If this product were a person, what would it be like? Male or female? Younger or older? What type of profession? What type of leisure activities?" All of these images affect how the role of the product is perceived and how it fits into the user's habits and lifestyle.

Product appearance and aroma tend to evoke very strong imagery associations and should be thoroughly explored before products are actually tasted. Appearance and aroma can convey appetizing or nonappetizing impressions regardless of the actual taste of a product. Therefore, it is critical that the projected imagery of these dimensions be consistent with the actual taste delivery.

J. Nonverbal Language

Observation of respondents' nonverbal language provides additional opportunities for assessing consumers' reactions. Subtle mannerisms such as nods of the head, eye contact, facial expressions, and loudness or quietness of the answers can provide important clues regarding the truth, conviction, and emotional content of answers. If a respondent verbally expresses a liking for a particular product but is shaking her head or gesturing negatively, the astute moderator can probe further to determine the reasons for the conflicting messages.

Sensitivity to the manner of speaking or inflections in the voice also give clues as to the sincerity and meaningfulness of responses. Especially as the respondent approaches fatigue, it is important to differentiate whether the respondent is just making up words to have something to say, or if the description is a legitimate expression of the way the respondent feels.

K. Presession Orientation and Diaries

Respondents can be told in advance of the group session to do some thinking about the category that will be discussed. At the time of recruiting, they can be asked to familiarize themselves with the different brands on the market and to become aware of their own preparation and usage patterns. Respondents can also be advised to keep a dairy noting their observations about the category and their habits, and to bring this diary to the group meeting. This technique ensures stronger involvement with the topics to be discussed and helps prepare respondents for the questions that they will be asked.

L. Show and Tell

Respondents can be asked to bring some of their own products from home to the group session. This technique is excellent for stimulating immediate discussion of the product category. The respondent's participation is made easier because their products can be used as props and references as they discuss their reactions. Respondents have something tangible and familiar to relate to and can make more specific comparisons if asked to make comparisons with new test products.

M. Preinterview Home Usage

Respondents can be given the test product to use in their homes before coming to the focus group interview. This increases their familiarity with the product's characteristics and enhances fluency during the questioning. This technique is particularly helpful when a very limited amount of test product is available for evaluation, and not enough is available for quantitative testing.

V. SELECTING RESPONDENTS

The choice of respondents comprising each focus group has a direct influence on the results. Adequate thought should be given to the criteria for screening eligible respondents. The objectives of the study will dictate what types of consumers are most appropriate. Examples of the types of screening requirements that might be considered are described below.

A. Product Usage

The most obvious people to be included in the study are those who would have the greatest interest in the product. These people are most likely to be currently buying a product similar or related to the one being tested. Inclusion of light, moderate, and heavy category users will provide broader feedback reflecting differing patterns of consumption and loyalty.

B. Nonusers

Nonusers who are positively disposed to the concept of the product can provide additional perspectives. They will provide insights into the product qualities they resist and will be more articulate about the qualities that remain to be satisfied.

C. Psychographic/Lifestyle Groups

Respondents can be screened to represent different lifestyle types, for example, busy professional singles who may be particularly receptive to certain fast food preparations or individuals high in health awareness who may be interested in more natural foods.

D. Verbal Fluency

Consumer language generation can be enhanced by ensuring that respondents have superior fluency with the English language. Screening exercises can differentiate respondents with a wide vocabulary from those whose vocabulary is relatively narrow.

For example, potential respondents can be asked to write as many words as they can using the letters "HCRAESER". Another exercise is for them to write as many words as they can beginning with the letter "B". In each case, the longer the list, the greater the verbal fluency.

E. Creative Ability

Consumers can be chosen for high creativity. This screening can be based on psychological tests of creative ability such as picture completion and unusual uses generation. In picture completion tests, a person is asked to add lines to incomplete figures to sketch interesting objects or pictures.

The unusual uses test requires the person to generate as many different uses as possible for a common object such as a brick or a paper clip. Subjects who display a strong ability to create novel pictures or uses can be expected to have more facility with generating lists of category descriptors, product associations, and new product ideas. Focus groups using creative consumers are sometimes referred to as supergroups.

F. Screening Out Professional Respondents

Professional respondents lie about their qualifications for a study in order to participate in the group and collect the cooperation fee. Such people will participate in focus groups repeatedly and will answer questions according to whatever they think the moderator wants to hear.

Since the comments from such respondents are not honest, the resulting information becomes invalid. To minimize use of professional respondents, all respondents should be rescreened before the interview. Key qualifications to be rechecked are nonparticipation in a focus group in the past 6 months (or 12 months, if required) and appropriate product category usage.

VI. CONSIDERATIONS FOR A SYSTEMATIC STUDY DESIGN

A. Standardization Across Groups

The procedure for presenting products for tasting should be standardized between groups. Each product should be subjected to the same type of evaluation under similar conditions. Order of product presentation should be systematically varied to minimize halo effect and order bias.

B. Controlling the Environment

The environment for the focus group should be free from any elements that may distort normal perception. For example, lighting should be well balanced and not impart a strange color to the food. The room should be well ventilated without pollutants from adjoining areas. For example, in mall settings a focus room could be too close to a restaurant, hair salon, or other highly aromatic environment.

C. Timing the Research

Ideally, optimal timing of the research is based on a rational consideration of normal daily consumption patterns, rather than upon convenience. In reality, however, practical limitations often prevent optimal scheduling.

If possible, interviewing should coincide with the time of day when the product would normally be consumed. Holding the research discussion in the same time context as normal usage ties the evaluation and feedback to feelings, memories, and associations that have been conditioned to this time period. These conditioned reactions may appear as sensitivities at one time of the day and for different attributes at another time. Top-of-mind associations for orange juice, for example, may be most prolific in the morning when more of it is consumed. Any product that is strongly skewed toward consumption at a particular time would benefit from normal usage scheduling or research.

D. Frames of Reference: Monadic and Comparative Evaluations

Consumer language generation is enhanced when respondents have the opportunity to analyze the product both on its own and in comparison to other products. Monadic, or singular, evaluations encourage the generation of uniquely characteristic descriptors. Comparisons of the product's similarities and differences relative to other standards can give better guidance in terms of the intensity of impressions and preferences. Respondents' comparisons can be made among current and test products, competitive products, an imaginary ideal product, or all of these.

When obtaining monadic reactions, it is advisable to emphasize that the product is to be judged on its own. The group can be told that they are to describe all the impressions they receive from this product by itself, not in comparison to anything else they may have evaluated during the discussion. This instruction needs to be repeated before each monadic exposure because the natural and easiest tendency for consumers is to describe how one product is more or less like the one before.

With comparative judgments, respondents must be advised that they must identify each comparison clearly so that the moderator knows which product is being referred to.

E. Cleansing the Palate

It is essential to remove all traces of flavor influences from one tasting before administering the next product. This may be done with a mouth rinse of water, a bite of unsalted cracker or white bread, or another neutral substance that will not impart its own taste. Respondents should be warned not to swallow too much water or eat too much cracker between tastings to avoid bloating their stomachs.

F. The Limits of Multiple Tastings

Generally, technical development would like to screen as large a number of prototypes as possible, often more than is realistically practical. Sensory overload and fatigue from multiple tastings will eventually decrease the respondent's ability to make accurate distinctions. Although the maximum number of tastings can vary with the type of product under evaluation, a safe rule of thumb is to limit the number to eight. If more products need to be evaluated, the presentations should be divided into sets of not more than eight, with a rest period of at least 20 min between sets.

VII. FRAMEWORK FOR ORGANIZING A CONSUMER LANGUAGE STUDY

There are two main approaches to obtaining consumer language. The first and most common method is the combination exploratory study. Here the study objectives concern multiple issues, and consumer language is obtained at the same time. The second approach is the pure language study where the focus is totally on generating and categorizing word lists.

A. The Combination Study: Language and Product Dynamics

The combination exploratory study is a practical initial step for delving into a new product category. This approach can be used to obtain an in-depth perspective of user needs and resistances within a category. It can also be used when a rough screening of prototypes is needed to reduce the number of a manageable level for quantitative testing. Consumer language exploration is then integrated into the discussion and a broad range of objectives can be accomplished. The combination approach is most often used because of practical considerations of time and cost. By using the same consumers, a large amount of information can be obtained in a short period of time and at less cost than addressing multiple issues with separate studies. The trade-off is in terms of session time available for covering a range of issues, and possible respondent fatigue.

The focus group discussion can begin by establishing the context of product usage. General usage questions may include:

1. What brands and types of foods are consumed in a category?
2. What is the frequency of usage?
3. How has usage changed over the years?
4. When and in what situations is the product usually used?
5. What things are added to this item before eating or drinking it?
6. For items requiring preparation, how is the food usually prepared, and what variations are tried?
7. What other foods can be substituted for this product?

Assessing consumers' feelings about a category can call attention to additional areas that can be probed. Such background information also sheds light on the relative importance of different product dimensions. This can provide guidance for additional attribute areas that can be probed for language development. General exploratory questions may include:

1. What are the general likes and dislikes in a category?

2. What are the main reasons for choosing products in the category?
3. What are the expectations for product performance?
4. How would consumers describe the ideal product?
5. How do they feel about the use or absence of certain ingredients?
6. What imagery do they identify with the product and what are the typical user associations?

After the necessary background information has been obtained, the language development questioning procedures can be applied.

B. The Pure Consumer Language Study

In a pure consumer language study, language generation is the primary goal. Here, the group discussion is not concerned with marketing- or advertising-oriented questions such as those related to product positioning. Rather, the focus is more technical. Here the concern is the sensory registrations that consumers make, and how they relate to known systematic changes in different stimulus products.

In the design of this type of study, products are systematically varied on singular dimensions. The main goal is to better understand what consumer-generated attributes mean in technical terms. By varying each stimulus product by only one dimension, the relation between specific product changes and how they are interpreted becomes more clear.

The procedure is very straightforward and is helpful for taking a fresh look at an existing product, and to make certain that all relevant attributes are covered in quantitative product evaluations. While the combination study tends to obtain reactions to a wider range of product types, the pure language study is more likely to focus on an individual product that is varied on key dimensions.

A beverage currently on the market can be studied to see what perceptions are associated with differing ingredient levels. Raising the sucrose level of a beverage, for example, may elicit comments such as "more sweet", "less bitter", "rich", "smoother", "too syrupy", and "more flavor". Lowering the sucrose level may elicit such responses as "watery", "no flavor", "bitter", and "not sweet enough". After consumer language is obtained for changes in one ingredient, the next ingredient change can be presented. Where possible, three levels of concentration (low, moderate, and high) of each ingredient should be used. This will provide better direction in terms of the preference for flavor intensities.

After all product descriptors have been recorded, responses can be grouped to show which attribute words correspond with which physical changes.

C. Language Development Procedure

The following outlines the product taste-testing sequence that might be used when evaluating six different products for language development in the beverage category. All answers are written down and then discussed.

1. Round One
    a. Product #1
       Describe the appearance.
       Describe the aroma.
       Describe the taste.
       Likes/dislikes/reasons why.
       How could the taste be improved?
    b. Product #2
       (Same as questions for Product #1)
    c. Compare Product #1 to Product #2 — describe the similarities and differences.
    d. Product #3
       (Same as questions for Product #1)

e. Compare Product #3 to Product #2 — describe the similarities and differences.

f. Compare Product #3 to Product #1 — describe the similarities and differences.

2. Round Two — repeat the identical procedure for products 4, 5, and 6.

D. Language Development Instructions

The respondents' tasks are in the following sequence. Respondents are to first notice the product's characteristics; second, write down these characteristics; and third, discuss their reactions and what they have written down.

1. Monadic Instructions

a. Do not taste the beverage yet. Take a good look at this beverage and just notice its appearance and color. Then, on the form next to looks, write all the words you would use to describe the way it looks to you. Please use very specific words. (Pause.)

b. Discuss looks (descriptions/likes/dislikes).

c. Now bring the cup closer to your nose and notice the aroma and smell of this beverage. On the form next to aroma, write all the words you would use to describe the way it smells to you. (Pause.)

d. Discuss aroma (descriptions/likes/dislikes).

e. Now take a sip and taste the beverage. On the form, next to taste, write all the words that come to your mind that describe the way it tastes to you. (Pause.)

f. Now, on the form, circle whether you "like" this beverage, "don't like" it, or are "neutral", and write down all the reasons why you feel that way about this beverage. (Pause.)

g. Discuss taste (descriptions/likes and why/dislikes and why).

h. Overall, thinking about your own taste preferences, how could the taste of this beverage be improved for you?

i. Now take a bite of cracker, and we'll:
    1. Try another beverage (or)
    2. Compare this beverage to product

2. Comparative Instructions

a. On the form, next to similarities and differences, write down all the ways products ____ and ____ are similar and all the ways that they are different. You can taste the two beverages again, but make sure that you take a bite of cracker between each tasting to clear your taste buds. Be as specific as possible, and please make sure that you identify which product code letter you are writing about when you describe the differences.

b. Discuss all the ways products ____ and ____ are similar to each other.

c. Now let's discuss all the ways that products ____ and ____ are different from each other. Refer to each product by letter code.

d. Now, take a bite of cracker and we'll:
    1. Try another beverage (or)
    2. Compare product ____ with product ____.

3. Language Development Form

Anything that makes the respondent's task easier and more organized will reduce confusion and optimize the productivity of the research. A customized form for recording responses that corresponds with the flow of the interviewing will help the respondents and simplify subsequent analysis. Figure 1 illustrates what might be used with the language development procedure.

VIII. APPLYING LANGUAGE FINDINGS TO QUANTITATIVE TESTING

Consumer language development can aid the product development process in two important ways. First, it provides a basis for understanding how consumers relate product dimensions and differences to personal perceptions. Second, it provides the spectrum of attribute descriptions for generating attribute lists that can be used for product tests.

Generally, if 50 or more people are thoroughly probed in the qualitative exploration, the researcher will have a good idea of the characteristics that form a common range of reactions in a category. Making a tally of responses obtained in the language development questionnaire form will generally provide enough important dimensions to create even the longest attribute lists for product testing.

If special testing issues require a more stringent language screening than can be based on qualitative judgment, an additional refining step may be desired. Data reduction techniques can be used with larger samples to limit the number of attributes, condense highly correlated attributes into singular dimensions, and identify particular attributes that are most sensitive to actual physical changes.

Once the range of important product dimensions has been established, a more standardized vocabulary can be used across product tests for a given category or subcategory. Undertaking a language development study offers both short-term and long-term benefits. The immediate benefit is the availability of a comprehensive set of relevant attributes for product guidance. In addition, the same attribute wordings can be applied to other corporate research such as attitude and usage tracking studies. Commonality of language allows for more possibilities of cross-study comparisons and thus effectively enlarges the overall data base.

Long-term benefits also are derived from standardization of consumer language. Having a comprehensive and consistent vocabulary makes historical comparisons possible and maximizes the usefulness of previous learning. Consumer language development provides a practical foundation for the exploration of new products. Consumer language research also can enhance the understanding of existing products within a changing competitive environment.

LANGUAGE DEVELOPMENT FORM

GROUP # _____   NAME _____

1. PRODUCT 1

    LOOKS: _____

    AROMA: _____

    TASTE: _____

    CIRCLE ONE: LIKE   DON'T LIKE   NEUTRAL

    REASON WHY: _____

2. PRODUCT 2

    LOOKS: _____

    AROMA: _____

    TASTE: _____

    CIRCLE ONE: LIKE   DON'T LIKE   NEUTRAL

    REASON WHY: _____

3. COMPARISON OF PRODUCT 1 AND PRODUCT 2

    SIMILARITIES: _____

    DIFFERENCES: _____

4. PRODUCT 3

    LOOKS: _____

    AROMA: _____

    TASTE: _____

    CIRCLE ONE: LIKE   DON'T LIKE   NEUTRAL

    REASON WHY: _____

5. COMPARISON OF PRODUCT 3 AND PRODUCT 2

    SIMILARITIES: _____

    DIFFERENCES: _____

6. COMPARISON OF PRODUCT 3 AND PRODUCT 1

    SIMILARITIES: _____

    DIFFERENCES: _____

7. PRODUCT 4

    LOOKS: _____

    AROMA: _____

    TASTE: _____

    CIRCLE ONE: LIKE   DON'T LIKE   NEUTRAL

    REASON WHY: _____

8. PRODUCT 5

    LOOKS: _____

    AROMA: _____

    TASTE: _____

    CIRCLE ONE: LIKE   DON'T LIKE   NEUTRAL

    REASON WHY: _____

9. COMPARISON OF PRODUCT 5 AND PRODUCT 4

    SIMILARITIES: _____

    DIFFERENCES: _____

10. PRODUCT 6

    LOOKS: _____

    AROMA: _____

    TASTE: _____

    CIRCLE ONE: LIKE   DON'T LIKE   NEUTRAL

    REASON WHY: _____

11. COMPARISON OF PRODUCT 6 AND PRODUCT 5

    SIMILARITIES: _____

    DIFFERENCES: _____

12. COMPARISON OF PRODUCT 6 AND PRODUCT 4

    SIMILARITIES: _____

    DIFFERENCES: _____

FIGURE 1.   Language development form.

Chapter 2

PROFILE METHODS:
FLAVOR PROFILE AND PROFILE ATTRIBUTE ANALYSIS

Anne J. Neilson, Virginia B. Ferguson, and David A. Kendall

TABLE OF CONTENTS

I. Introduction 22
    A. Flavor Profile 22
    B. Profile Attribute Analysis 23

II. Dynamics of Profiling 23
    A. Basic Principles 23
        1. Psychophysical Issues 24
        2. Social Psychological Issues 24
        3. The Training Issue 25
        4. The Suggestion Issue 25
    B. Procedures for Panel Operations 26
        1. Orientation 26
        2. Formal Sessions 26
        3. Operating Conditions 26
        4. Standardized Smelling and Tasting Techniques 27
        5. Reference Standards 27

III. Flavor Profile 28
    A. Definition of Profile Terms 28
        1. Character Notes 28
        2. Intensity 28
        3. Order of Perception or Order of Appearance 29
        4. Aftertaste 29
        5. Amplitude 29
    B. Establishment of a Final Flavor Profile 29
    C. Reporting Flavor Profile Information 29
    D. Statistical Treatment of Flavor Profile Data 31
    E. Applications 31
        1. Uses of the Flavor Profile 31
        2. Examples 31
            a. Ross Laboratories (A Division of Abbott
                Laboratories) 31
            b. Metropolitan Water District of Southern
                California 32

IV. Profile Attribute Analysis 32
    A. Procedures for Panel Operation 32
    B. Selection of Profile Attributes 33
    C. Attribute Scales 34
    D. Statistical Procedures 36
    E. Advantages of PAA 37
    F. Uses of PAA 37

V.   Panel Training Programs.. 37
     A.   Selection of Trainees 38
          1.   Criteria... 38
          2.   Screening Tests 38
     B.   Panel Training Courses.. 39
          1.   Introductory Course... 39
          2.   Advanced Course 39
          3.   Monitored Work Program Assignments.............................. 39
          4.   Post-Instructional Guidance and Follow-Up Visits 40
          5.   Panel Leader Recommendations and Certification
               of Trainees.. 40
          6.   Implementation 40

References.. 40

I. INTRODUCTION

The two profile methods of sensory analysis are Flavor Profile and Profile Attribute Analysis. Both were developed by Arthur D. Little, Inc., Cambridge, Mass., and both make use of trained flavor panelists.[1,16] The Flavor Profile Method is highly descriptive. Emphasis is placed on analysis of one product at a time to obtain a detailed outline of its aroma, flavor, and aftertaste characteristics. It is an excellent method for use in product definition and prototype development.

Profile Attribute Analysis is both a descriptive and numerically based method of sensory analysis. This method differs from Flavor Profile in that more than one sample is analyzed at one time, and the findings can be treated statistically. This method is suitable for use when many variables are under study and/or a large number of samples must be evaluated in a relatively short time.

A. Flavor Profile

Arthur D. Little, Inc. developed the Flavor Profile Method in 1949. This outline presents the latest refinements in the method based on its successful use worldwide since its inception. Previous authors have discussed the nature of the Flavor Profile,[1] the general organization and operation of profile panels,[2,3] and the basic method of training panel members.[4] All of these aspects are summarized herein.

The Flavor Profile Method is empirically based, i.e., it was developed and learned through experience and employs perceptual judgments of both the elements and structure of aroma and flavor impressions. These judgments are made by carefully selected and extensively trained panelists who work as a team to reach a concensus. Statistics may be employed to analyze Flavor Profile data, although reproducibility of profile results is generally based upon the skill and training of the individual, and the use of objective reference materials to eliminate discrepancies in terminology or scaling.

Two years of application studies, directed in particular toward explaining the changes in sensory perceptions caused by seasoning cooked foods, lay behind the development of this method. Prior to the use of a systemized approach, it was impossible to describe accurately the sensory effects of adding materials such as salt, pepper, butter, and monosodium glutamate (MSG).

Since its development, the Flavor Profile Method has been successfully applied to many problems involving food, beverages, pharmaceuticals, tobacco products, packaging materials, textiles, household products, and toiletries. It has proven to be a valuable tool for research, product development, and quality control, presenting complete descriptions of the perceptual attributes of each product analyzed. This is possible because, in addition to word descriptors, the method employs intensity scaling for each sensory impression and ratings to indicate the degree of blend and fullness of the aroma and flavor.

Interpretation of panel findings involves comparison of a particular sample's Flavor Profile with Flavor Profiles of the other samples in the set under evaluation. This form of interpretation has several advantages: it provides specific descriptions of the individual sensory impressions in aroma, flavor-by-mouth, and aftertaste; it indicates the order of appearance and changes in that order; and it gives intensities or indications of perceptual strength for each word descriptor.

All of this information can be integrated when the total sample size is relatively small, but the process becomes unwieldy when the study involves hundreds of samples and large volumes of Flavor Profile data. At these high sampling levels, two serious problems become evident: it is difficult to isolate those variables that indicate significant process or product changes, and the logistics and costs of collecting the data become prohibitive.

These problems led Arthur D. Little, Inc., to develop a more cost-effective method of sensory analysis, a system called Profile Attribute Analysis (PAA).

B. Profile Attribute Analysis

PAA is an extension of the Flavor Profile Method in concept and implementation. As does the Flavor Profile Method, PAA uses trained panelists to define and measure the attributes necessary to differentiate a set of samples in a given product category. A small group of four to six trained panelists can produce sufficiently uniform results to provide significant discrimination among samples after two or three replications.

PAA provides an objective method of describing numerically the complete sensory experience, i.e., the appearance, texture, aroma, flavor, and aftertaste. By limiting the number of sensory attributes which the panelists measure, it is possible to evaluate as many as five samples in the time allowed for the more detailed Flavor Profile description of only one sample, making PAA more cost-effective and efficient. Because the data generated are numerical, they are efficiently stored in automated data handling systems and the most modern, robust statistical techniques are applied to summarize the data and assist in the interpretation.

The Flavor Profile Method introduced the concept of an integrative perceptual measure of aroma and flavor. This measure includes the sensations of balance and fullness, which are perceptions used not only by experts, but also by consumers, to distinguish among samples. PAA uses both balance and fullness as independent measures.

Finally, where Flavor Profile describes sensory characteristics in terms of three chemical sensory pathways (smell, taste, and mouthfeel), PAA adds visual and tactile measures to this list and occasionally includes auditory perceptions, such as the snap of a potato chip, to describe completely the products of concern.

II. DYNAMICS OF PROFILING

A. Basic Principles

Perception has been defined as the integration of various stimuli or sensations into a complex whole, as well as the isolation of elements within the complex whole. The concept of perception was first introduced early in the 20th century by the German psychologists who called themselves "Gestaltists". This group was principally interested in visual perception, and their research was directed toward demonstrating that visual perception is a complex process which involves selecting and organizing individual "sensations" into complex patterns that are more than a summation of the individual sensations themselves. They maintained that what you "see" cannot be predicted by a simple analysis of individual sensations because these sensations interact to produce patterns or configurations which are not properties of the individual sensations themselves, but rather of the way they are arranged or organized.

Thus, perception differs from simple sensory measurement in many ways. When we look at a painting, we usually do not notice the individual brushstrokes; rather we see an integrated whole. Listening to music, we are less aware of the individual notes then we are of a melody. Of course, we can look beyond the whole and analyze individual elements, such as brushstrokes or individual notes, and these tell us something about how to build the whole, but the initial and overriding perception is one of the totality created by the arrangement or organization of the elements.

It can be said that flavor is to taste as perception is to sensory measurement. Profile Methods deal with perception, whereas other techniques deal with taste measurement.

1. Psychophysical Issues

Psychologists classify the Profile Methods as phenomenological as opposed to psychometric or psychophysical. Psychometric taste tests are exemplified by difference tests which determine the number of persons finding a difference between samples. Psychophysical tests measure the intensity of a stimulus or combination of stimuli perceived by trained or untrained subjects. Early Flavor Profile researchers found that neither psychometric nor psychophysical tests provided all the information needed to define the complexity of perceptual change that takes place when food products are fermented, cooked, or seasoned. These other tests also are inadequate for solving flavor problems, for example, describing the effects on blend, fullness, bitterness, and aftertaste produced by artificial sweeteners.

Profile Methods measure both the overall impression created by a flavor or odor and the individual elements which contribute to that impression. The Flavor Profile Method introduced the concept of overall impression to flavor evaluation and termed it amplitude. Amplitude is a measure of the degree to which all aspects of an aroma or a flavor are blended together to form a whole. Amplitude is derived from the word *ampleness*, an indication of fullness. For example, a sugar-sweetened cola has noticeably more blend and fullness than a diet cola with saccharin.

Inclusion of the amplitude concept in sensory measurement methods not only has relevance to Gestalt psychology, but also to modern perceptual psychology and recent developments in the psychology of consciousness and human perception.[5] For example, evidence shows that the two hemispheres of the brain are specialized in their functions. The left, or so-called "dominant" hemisphere, is analytic and operates according to the traditional definition of rational "scientific" thinking. The right hemisphere, on the other hand, is nonrational, intuitive, and integrative, functioning in a more "artistic" manner. In other words, the left hemisphere analyzes, while the right synthesizes. Thus, since analysis and synthesis are inherent in the perception process, both are incorporated into Profile Methods. Experience has taught that flavor panels comprised of individuals who only see the trees (left brain) rather than the "forest" (right brain) often miss a significant part of the total sensory impression.

2. Social Psychological Issues

Profile Methods are based on the use of a panel of four to six persons who have normal abilities of smell and taste, are trained in techniques of smelling and tasting, have had experience as panel members, and undergo a training period to orient them to the particular product to be analyzed. One of the panel members also acts as the panel leader.

The selection of a panel leader is based on several criteria. Most important is the candidate's demonstration of excellent communication skills, both oral and written. This includes eliciting information from other panel members, compositing and reporting results, and acting as liaison between the panel and those requesting sensory information. Another requirement for panel leadership is the ability to motivate the panel in order to maintain the level of interest and skill necessary for optimum performance. The panel leader has to maintain objectivity in evaluating and reporting sensory data, and exhibit a firm, effective, but nondomineering manner in all facets of panel operation.

However, this "leader" does not act as a superior in any way, but is simply the panelist responsible for coordinating the activity of the group and reporting its results. The leader does not necessarily have any special knowledge, and any panelist theoretically can function in the leadership role.

This is in contrast to sensory measurement methods which employ the traditional subject-experimenter role relationship. In such panels the subject (panelist) is treated in a formal and impersonal manner as a subordinate, is given explicit instructions by the experimenter, and is not expected to exhibit any individual initiative.

Experiments have shown that an "autocratic" style (as in the traditional experimenter role) produces a group behavior which can be irresponsible and lacking in initiative, and in which the members experience very little enjoyment in carrying out their tasks and manifest hostility toward both their autocratic leader and toward one another.[6] A democratic style of leadership, on the other hand, in which the leader merely facilitates group activity and decision making, has been found to foster cooperation, synergy, and a marked increase in productivity. The profile approaches thus rest upon role relationships which are models of those contemporary social psychologists advocate in the place of the traditional subject-experimenter relationship.

3. The Training Issue

The training of panelists has always been an important aspect of Profile Methods. This training is based on the thesis, supported by evidence from psychological research, that individual stimuli are much more accurately perceived when the observer is familiar with the kinds of stimuli for which he or she should be looking.[7] In other words, an untrained observer is likely to miss a great deal. In addition, the observer also must have the appropriate vocabulary to label or describe these perceptions.

Thus, the Profile approaches involve detailed education of panelists. This training (see Section V) teaches participants an extensive vocabulary, the intensity scale, and the techniques of smelling and tasting.

4. The Suggestion Issue

Suggestion is an important issue in the psychology of perception and cognition. The need to guard against the so-called "experimenter expectancy effect" is one of the central concerns of all psychologists conducting research with human subjects.[8]

An important aspect of Profile Methods is the use of a blank sheet of paper at all orientation and initial panel sessions, instead of a "check list" as is commonly employed in other methods of sensory analysis. A check list is by its nature suggestive, implying what the experimenter expects to be perceived and frequently even the order of perception. The "blank sheet" format eliminates the suggestive nature of a check list, limiting panelists' judgments to what they perceive in the product. For example, panelists are reported to be more inclined to report unusual character notes when their perceptions are not artificially narrowed or limited by a predetermined list of descriptors. The blank sheet also allows panelists to list character notes as they appear, not influenced by the order in which preselected characteristics are listed on a page in front of them. The blank sheet format is used at the initial panel session or sessions for each study. Subsequent panel sessions, which are required to develop a final profile, may make use of a reponse sheet summarizing the initial information. Panelists are encouraged to make as many changes as necessary to describe the products adequately.

B. Procedures for Panel Operations

Profile panels work in a standardized fashion and under conditions conducive to concentration. Flavor is at best elusive, so adequate time and space and the absence of distracting noise and atmosphere are necessary for the proper functioning of the panel. Also, the need for highly standardized techniques of smelling and tasting and standardized methods of preparing and presenting samples cannot be overemphasized. More specifically, five elements are prerequisites to good analytical work:

1.   Orientation
2.   Formal sessions
3.   Proper room conditions
4.   Standardized smelling and tasting techniques
5.   Reference standards

1. Orientation

Preparation for formal profile panels requires a period of orientation. A trained or experienced panel will usually require at least one to two informal sessions, and a less experienced panel may require several more. The general procedure involves a meeting of the panel where the panel leader outlines the objectives of the project. Even experienced flavorists need orientation to a specific problem in order to work as a unit toward a particular objective, whether this be characterization of an off odor or flavor, comparison of a prototype with a market sample, or maintenance of quality control. The best conditions for handling and presentation of samples, including temperature control, are also decided upon at this time.

However, the primary focus of the orientation sessions is to develop a descriptive vocabulary for use at subsequent sessions. Only when all panelists believe that the orientation is complete is the panel ready to proceed with formal flavor panel sessions.

2. Formal Sessions

Formal profile sessions consist of a closed panel followed by an open panel. In the closed panel, each participant makes an independent evaluation of the products presented. In the subsequent open panel, each panelist recites his or her findings in an open discussion led by the panel leader, who also compiles the findings. The oral review of findings gives the panel members immediate feedback as to their ability to describe what they have perceived and indicates where practice may be necessary. This session centers attention on the products under study, clarifies terms, and enables the panel to select suitable reference standards. Open discussion is an integral part of profile methods where experiences are exchangeable and additive. It helps trainees to become sound panel members and assists panel members in learning the techniques of leading the panel.

3. Operating Conditions

A profile panel acts as an analytical instrument, so control must be maintained over extraneous variables. A panel works in a well-lighted, clean, quiet, odor-free, temperature-controlled room equipped with a round table and chairs.

Panel sessions are held at regularly scheduled times to maintain an orderly progression in the study. Panelists are encouraged to attend as many sessions as possible to keep their skills sharp. Interruptions are not permitted during the sessions. The samples to be profiled should be as uniform as possible, and each panel member should receive an aliquot of the sample for his or her independent analysis.

The flavor analyst should be allowed to concentrate entirely on the product being studied. The plates and utensils used are not only clean but odor-free. Because flavor properties differ at different temperatures, the temperatures of all panel members' samples must be the same and be consistant from day to day. Furthermore, all panel members must examine their samples in the same way: all take the same number of sniffs for aroma and the same number of the same size bites or sips for flavor. All measure aftertaste at the same number of minutes after the last swallow.

Each member follows a standardized procedure that the panel has adopted to achieve the most useful and complete analysis of the particular product under analysis. For example, with iced beverages, the number of ice cubes and the time of drinking after adding the ice must be specified.

The number of panel sessions necessary for completing profile analyses varies according to the complexity of the product. Customarily, final flavor profiles can be generated after three or four sessions.

4. Standardized Smelling and Tasting Techniques

A product's aroma is analyzed first, then the flavor, and finally the aftertaste, which constitutes the sensations still present in the mouth 1 min (or other designated time) after swallowing.

Aroma is defined as the sensation perceived by the nose when the product is sniffed. It is composed of volatile components (such as the sweet fragrance of vanilla) and feeling sensations (such as the coolness of menthol). Generally, a sample should not be handled while smelling because of possible interference from odor which may be on the hands, or from temperature change introduced by touching the dish or glass. Instead, the panelist brings his or her head down to the sample and takes a few short "bunny-like" sniffs. Small sniffs repeated three times are all that is needed to perceive and describe the aroma sensations. This limitation avoids saturation and fatigue of the sensory organs.

Exhaling on dry products such as cold cereal or crackers presented in a dish or on a plate is sometimes a useful tool to increase the level of aromatics. Notation should be made that this technique was used, and all panelists should follow the same procedure.

Flavor or *flavor-by-mouth* is perceived when the sample is taken into the mouth and swallowed. The term flavor-by-mouth is used to indicate that the evaluation was made through tasting and not by eye or by ear.

Flavor is a combination of aromatics or the volatile components which reach the olfactory area, basic tastes (i.e., sweet, sour, salty, and bitter), and feeling factors such as pepper bite, mustard burn, or lemon juice astringency. Panelists should always "taste" and not eat or drink as too much quantity can lead to saturation of the taste buds. Small sips (5 to 10 mℓ) are adequate to cover the entire tongue, although liquids may also be slurped so that the aromatics can reach the olfactory area.

5. Reference Standards

The use of the same term to describe a given sensation is absolutely essential in Profile Methods. This is attained first through general training in odor and flavor perception, and then through specific practice with the product at hand. Each term listed in a profile tabulation is understood by all the panel members and can be illustrated through the use of a reference material. Basic tastes are generally referenced by aqueous solutions of sucrose (sweet), citric acid (sour), sodium chloride (salty), and caffeine (bitter). References for aromatics can be essential oils, natural plant materials, spices, herbs, or flavorings such as vanilla and almond extracts. Chemicals such as menthol (cool and burn), capsaicin (bite), alum (astringent), and calcium carbonate (chalky) in aqueous solution provide good reference standards for feeling factors.

Reference standards are also used for intensity calibration. These standards are most commonly solutions of different concentrations of the basic tastes.

III. FLAVOR PROFILE

A. Definition of Profile Terms

Each panel member looks for the following flavor characteristics when doing a flavor profile:

1.   Character notes
2.   Intensity
3.   Order of perception or appearance
4.   Aftertaste
5.   Amplitude

In addition, when texture and color are important to the product's description, they are also noted during the panel session.

1. Character Notes

The individual components of aroma and flavor are called "character notes" and are defined in descriptive or associative terms. In the Flavor Profile Method, each product category has an associated glossary of terms with which all panelists are familiar. These terms are objective rather than subjective (e.g., a character note might be labeled "vanilla" but not "good") and easily referenced. As panelists become experienced, they can quickly build a vocabulary of 300 to 500 such descriptors.

Note that in some cases, some panelists will describe a character note that the others do not perceive. In cases where less than half of the panel members perceive a note, it is referred to as an "other" and listed at the bottom of the flavor profile.

2. Intensity

The degree to which each character note is perceived is called "intensity." The intensity scale is simple and constant:

0   Not present
)(Threshold or just recognizable
1   Slightly strong
2   Moderately strong
3   Strong

Since these are fairly broad demarcations, experienced panelists will want to designate narrower ranges. They do this by using symbols such as 1/2, 1 1/2, or 2 1/2.

Threshold, signified by the symbol)(, is really a statistic rather than one definite point. It represents the range of concentration of a material's aroma or flavor that is detected 50% of the time by the panelists.

The intensity scale is used for all product categories. When sweet taste is perceived by the panelist as being moderate in intensity, it is rated by the symbol 2, whether the panelist is measuring coffee with two teaspoons of sugar, a carbonated cola beverage, a sweet wine, or a fresh fruit.

3. Order of Perception or Order of Appearance

In addition to giving objective descriptions of the perceptible factors and their intensities, a tabular profile lists these factors (character notes) in the order in which they are perceived. This is made possible by the use of standardized techniques of smelling and tasting as order of appearance is influenced by the perception of basic tastes on the tongue and perception of volatiles in the nose. Texture and blend also influence order of appearance. Differences in the time of appearance of character notes are more apparent in unblended flavors.

4. Aftertaste

Aftertaste or aftereffect is a definite and important part of flavor. The panelists note and report these impressions at a predetermined time after completion of tasting. Aftertaste sensations can include basic tastes, aromatics, feeling factors, or all three impressions. Generally intensity ratings are not used, but for specific studies where aftertaste is important, intensity ratings add further definition.

5. Amplitude

In doing a flavor profile, panel members rate the amplitude — the degree of blend and the amount of fullness present in the aroma and the flavor — before they concentrate on the individually detectable character notes of the product.

Amplitude is rated on a seven-point scale composed of four major ratings with three intermediate (1/2) ratings between them:

0 = no blending and fullness
1 = low degree of blend and fullness
2 = moderate degree of blend and fullness
3 = high degree of blend and fullness

B. Establishment of a Final Flavor Profile

Flavor Profile panelists are first exposed to the sensory impressions of a product under study in the orientation session. At the conclusion of the first panel session a preliminary flavor profile is generated by the panel leader. This profile is continually refined during subsequent sessions until all panelists agree that a final composite judgment has been reached. This then becomes the final Flavor Profile. Generally, it takes three to five sessions to produce a final composite.

Table 1 provides an example of the format used to write a Flavor Profile. Table 2 illustrates a means for comparing two flavor profiles.

C. Reporting Flavor Profile Information

The Flavor Profile report should include complete identification of the sample(s) studied, and the objectives and duration of the study.

The report presents data which can be used to support recommendations of product or process changes, modification of quality-control or shelf-life policies, etc. It should be as concise and brief as possible.

The body of the report should include the techniques used to examine the products, such as preparation methods and serving temperatures, so that an evaluation could be repeated accurately in the future. The report should also identify reference standards used to reach agreement on descriptive terminology and intensities. A tabular presentation of the profile data (see Table 1) is not always submitted, but when it is, it must be accompanied by written

Table 1
FLAVOR PROFILE FORMAT

Aroma
    Amplitude   Rating
    Character note   Intensity (in order of appearance)
    Etc.
    Others:
Flavor
    Amplitude   Rating
    Character note   Intensity (in order of appearance)
    Etc.
    Others:
Aftertaste
    Character notes (intensities optional)
Footnotes (optional)
    Color:
    Texture:

                                                ——————————————————
                                              Name and signature of panel leader

                                                ——————————————————
                                                Date submitted

Table 2
EXAMPLE OF FLAVOR PROFILES OF
MAYONNAISES

Modern® Mayonnaise		Yankee® Mayonnaise	
Code: September 28, 1984	**Rating**	**Code: August 9, 1984**	**Rating**
Aroma Amplitude 2		**Aroma Amplitude 1$^1/_2$**	
Oily, vegetable	1$^1/_2$	Sour, vinegar	2
Eggy, cooked	1	Oily, slightly oxidized	1$^1/_2$
Sour, vinegar	1$^1/_2$	Pungent	1
Spice complex garlic, mustard	$^1/_2$	Spice complex garlic, onion	1
Other: Briny		Other: Eggy	
Flavor		**Flavor**	
Amplitude 2		**Amplitude 1**	
Sweet	$^1/_2$	Sweet	1$^1/_2$
Oily, vegetable	1$^1/_2$	Sour, vinegar	2$^1/_2$
Salty	2	Oily with mouthfeel	1
Sour, vinegar	1$^1/_2$	Salty	1
Eggy, cooked	1	Spice bite and burn	1
Oily mouthfeel	1$^1/_2$	Astringent	1$^1/_2$
Other: Spicy		Other: Eggy	
Aftertaste		**Aftertaste**	
Salty		Sour	
Oily with mouthfeel		Spice burn	
Color		**Color**	
Slight eggy yellow		Pale creamy white	
Texture		**Texture**	
Smooth, gelatinous		Lumpy, grainy	

discussion to aid interpretation of the information. Regardless of whether or not the tabular profiles are included, the report should contain a summary of the amplitude ratings, major character notes of aroma and flavor, and their intensities. It also should mention the presence or absence of off-notes, and discuss order of appearance and aftertaste. Observations of visual and textural qualities are also important.

Frequently organizations adapt a standard format for reporting flavor profile information which provides a ready reference for action. Also, many companies store flavor profiles on microfiche or in computer data bases.

D. Statistical Treatment of Flavor Profile Data

The Flavor Profile Method was not designed as a numerical system to allow for statistical analysis of the data. Most applications are based on interpretation of the composite flavor profile terms and intensities. However, the Flavor Profile does provide multivariate data which can be summarized to provide a statistically treatable comparison of several like products.

This process involves making a transformation to an expanded scale by assigning numbers to an aroma note index and flavor note index, summarizing the indexes, and using Principal Component Analysis and Analysis of Variance to treat the data. A case study where this was done for fluid milk products is discussed in detail by Dr. Irwin Miller.[9] Other studies employing statistics have been reported in the literature.[10]

E. Applications

1. Uses of the Flavor Profile

A Flavor Profile panel can serve many valuable functions. It can guide the development of new flavor prototypes, provide a permanent record of the aroma and flavor of a product at given points in time, and measure changes caused by storage or packaging. In the same manner, the panel can contribute to product improvement by studying ingredient and processing variables. It can also guide quality control of both raw materials and finished products. As flavor profile data provide a company with a record of the sensory qualities of its products and those of its competitors, the information is very useful to consumer product testing. Prior to a test, Flavor Profile panels can be used to screen prototypes for selection of those suitable for testing. Profile data are helpful in the design of questionnaires and in interpreting consumer response. It is wise, too, to have a Flavor Profile panel monitor the products under test during the time of the test to ensure that no significant or unexpected changes in sensory impression have taken place.

The ways in which trained and experienced Flavor Profile panels are used are as diverse as the industries using them. Many food companies use Flavor Profile panels, but as the two examples below illustrate, there are less conventional applications as well.

2. Examples

a. Ross Laboratories (A Division of Abbott Laboratories)

Ross Laboratories considers flavor the most important attribute in the marketing of their medical nutritional products, because flavor is critical to patient acceptance of oral nutritionals.[11] In recognition of this, Ross has three sensory evaluation panels. Each panel is composed of Ross employees with different degrees of sensory training. One of these is a Flavor Profile panel which prepares objective analyses of the aroma, flavor, and aftertaste of Ross Laboratories' products. This information is used to select types and levels of flavoring ingredients to improve palatability, to measure the effects on aroma and flavor of various packaging materials, to reduce costs and increase ease of use, and finally, to guide research in extending high-quality shelf life.

b. Metropolitan Water District of Southern California

At Metropolitan,[12] the traditional method of determining water quality, a threshold odor number (TON), often failed to provide sensory data of reliable use except in cases of gross contamination. Therefore, a Flavor Profile panel was established to provide Metropolitan with an effective sensory technique for evaluation of drinking water.

The Flavor Profile panels have been used to monitor earthy flavors produced by micro-organisms in sourcewater reservoirs, to monitor medicinal flavors caused by phenolic compounds, and to assist in controlling problems with oil seepage into a raw-water tunnel and solvents leaching into a newly covered reservoir.

Flavor Profile analysis has provided the water quality management with reproducible sensory data and information on the presence of numerous organic compounds in the water. The results have been produced in an expeditious manner and have precluded certain costly gas chromatograph/mass spectrograph analyses.[13]

IV. PROFILE ATTRIBUTE ANALYSIS

PAA is an objective method of sensory analysis that uses an expert panel to describe numerically the complete sensory experience of flavor through profile attributes. These attributes consist of a limited set of characteristics which, when properly selected and defined, provide a meaningful description of the sample. Additional detail is provided by panel comments which are stored in association with the attribute data. By limiting the number of profile attributes the panelist measures, it is possible to evaluate five samples per session using PAA, compared to one sample per session using the Flavor Profile Method. PAA data can be efficiently stored in automated data handling systems and are amenable to statistical analysis and data summarization.

PAA is based on the Flavor Profile Method in concept and implementation. It uses the expertise of trained panelists to measure the critical set of attributes for a specific study. A group of four or five trained panelists, drawn from a pool of six to eight who are familiar with the specific study, produces sufficiently uniform results so that two or three replications can provide significant discrimination among all of the samples in the study.

The Flavor Profile Method introduced the concept of integrative perceptual measures of flavor in the amplitude characteristic, which includes aspects of balance and fullness. PAA uses both of these terms as independent measures.

The Flavor Profile Method describes character notes in three chemical sensory pathways — smell, taste, and mouthfeel. PAA adds visual and tactile measurements to this list. These added sensations provide additional ways of differentiating among samples both by consumers and by experts.[14] Further, as visual and tactile modalities frequently alter or modify one's perceptions of smell, taste and mouthfeel, it is important to be able to measure their interactive effects.

A. Procedures for Panel Operation

PAA panels are held in panel rooms free from extraneous odor and from interruption. The sessions are scheduled at regular intervals to prevent panelist fatigue from too much tasting, as well as to reinforce the discipline required for effective PAA performance. Standardized smelling and tasting techniques are used for all products.

Prior to conducting formal PAA sessions, orientation meetings are held to establish the attributes and rating scales for the products to be analyzed and to prepare a response sheet for subsequent use. Frequently, samples will be flavor-profiled prior to selecting the PAA attributes. It is important that a proper experimental design be established prior to initiating the study. A matrix based on brands, flavor type, package, plant location, and process will generally be adequate for most studies.

At the formal PAA sessions, each panelist is presented with a number, typically five, of unidentified samples for evaluation. The panelist tastes and scores each sample on the response sheet provided. Three small tastes, approximately 1/2-in. cubes for solids or 5 to 10 mℓ for liquids, are sufficient. One minute after swallowing, residual flavors are scored as aftertaste. The next sample is tasted in a similar fashion after waiting another minute to assure that carry-over will not influence response. Analysis of five samples usually requires 15 to 20 min of concentrated effort.

The scores and comments on each sample are then collected from each panelist using a data control sheet and concurrently entered into the data management system. Descriptive comments are discussed and a consensus recorded. A composite score for each attribute is announced so that individuals can determine how their responses compare with the group consensus. Discussions at this time help panelists identify areas of disagreement, and also help ensure that panelists are calibrated to the attribute scales. If the scores for any sample diverge more than two units, the analysis is repeated.

Each sample is examined at least twice. Products such as cigarettes and distilled spirits, which are difficult to analyze, sometimes require three or four replications to ensure an accurate analysis of the sample. Throughout each study, the samples are presented randomly to prevent ordering bias.

The scores for all attributes of each sample are entered into an appropriately programmed data file by a panelist. The composite scores are also stored in the file for comparison with the average results. The computer system is used to average the scores for each sample and to print the results of the panel session. Replicate analyses are printed individually and combined to produce an average sample result.

B. Selection of Profile Attributes

Attributes are selected at the orientation sessions. At this time the panelists taste and smell a wide variety of products within the product category to be studied. Generally, the categories are broad. For example, for a cookie improvement study, approximately 50 different cookies, spanning textural differences from hard to soft and flavor differences from chocolate to raspberry, would be required to reflect the possible variations adequately.

At least two samples that represent extremes of flavor variation are characterized using Flavor Profile Analysis. These data are compared to note the sensory impressions that seem to reflect the differences between the samples. The differences are listed so that at subsequent orientation sessions the panelists can group them into attributes to provide the most complete means of differentiating between the samples.

Experience has taught that integrative perceptual attributes are an essential part of a product description. The attributes of balance and fullness not only distinguish among samples, but they are closely correlated with consumers' descriptions of the same products. It may also be necessary to include other integrative measures such as total intensity of flavor (or impact), complexity of aromatics, and aftertaste to differentiate the samples fully in specific situations.

The analytic attributes include measurement of color, texture, type of aromatic identity, basic tastes, and specific mouthfeel sensations. Combining these lists would provide the following attributes for the hypothetical cookie study:

Attributes describing integrative perceptions	Attributes describing analytic perceptions
Balance	Color
Fullness	Texture
Aftertaste	Aromatic identity
Others	Sweet taste intensity
	Mouthfeel

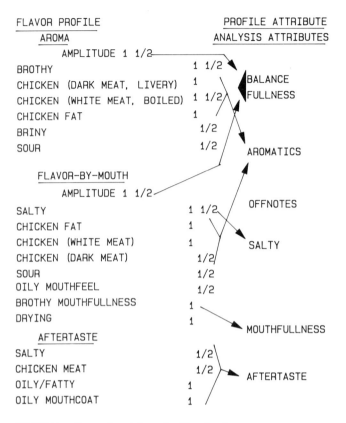

FIGURE 1. Conversion of Flavor Profile to Profile Attributes — chicken broth with 0.5% sodium chloride.

In different situations PAA can be used to measure prototype or experimental samples against a standard. For example, the development of a salt substitute can be guided by measuring the effect of laboratory samples in a simple medium, such as chicken broth, and scoring these samples on attribute scales developed for chicken broth with sodium chloride. Figure 1 illustrates the process of converting the flavor profile of chicken broth with 0.5% added sodium chloride to PAA attributes.

C. Attribute Scales

Once the profile attributes are selected, the attribute scales are defined. The panelists are made familiar with the definition and anchor points through the use of reference standards. All attributes are defined along an integer scale of 1 to 7, with each unit representing a difference the panelists can detect with confidence on repeated analyses.

For attributes such as aromatics and mouthfeel, where learning the strength of the sensory impressions is desired, the 1-to-7 PAA scale is used to represent flavor profile intensity ratings, as follows:

PAA score	Flavor profile intensity
1	0
3	1 (slight)
5	2 (moderate)
7	3 (strong)

Sodium chloride

Attributes	Flavor profile	PAA score	Experimental samples PAA score		
Balance		4	Balanced		Unbalanced
	Amplitude 1½		1 2 3 4 5 6 7		
Fullness		3	Full		Thin
			1 2 3 4 5 6 7		
Salty	Intensity 1½	1	Typical	Atypical	
			1 2 3 4 5 6 7		
Aromatics					
Chicken dark meat	½	1	Typical	Atypical	
Chicken white meat	1		1 2 3 4 5 6 7		
Chicken fat	1				
Mouthfulness	1	1	Typical	Atypical	
			1 2 3 4 5 6 7		
Off notes	0	1	None		Many
Type			1 2 3 4 5 6 7		
Aftertaste	Slight	2	Short		Long
			1 2 3 4 5 6 7		

FIGURE 2.   Profile Attribute Analysis flavor response sheet—chicken broth with salty compounds.

With the integrative attributes, the scales reflect degrees of effect. For example, with fullness, the following numbers are used:

Revised
as of August 8, 1986

PAA score	Flavor profile term
1	Very full
4	Moderately full
7	Very thin

It is important that all panelists agree upon and understand the anchor points of the scale for every attribute. Generally, it is easy to obtain good reference standards for attributes such as color, texture, and appearance by using products within the category. For example, soda water and root beer provide good color endpoints; airy cheese snacks and beer nuts are good for a texture attribute, and different concentrations of carboxymethyl cellulose can be used to illustrate different perceptions of viscosity. In cases where products are measured against a standard, as in the salt substitute example, the standard generally scores a 1 for each analytic attribute. The integrative attributes of the standard (balance, fullness, and aftertaste) are scored according to the flavor profile. The total scores for the experimental samples then express the degree of closeness to or difference from the standard (see Figure 2).

In cases where it is possible to taste the standard, it is included among the samples. In this case the chicken broth with salt would be scored at 1 for most attributes. Note that the aromatics and off-note attributes include a rating of type of sensory impression as well as of intensity.

As all comments relating to the type of effect are included with the numerical attribute ratings, the data summaries provide quite explicit direction for the product development team. Attributes can also be derived from focus groups and/or consumer testing. It is important that each attribute be defined objectively and that suitable references are selected for the anchor points.

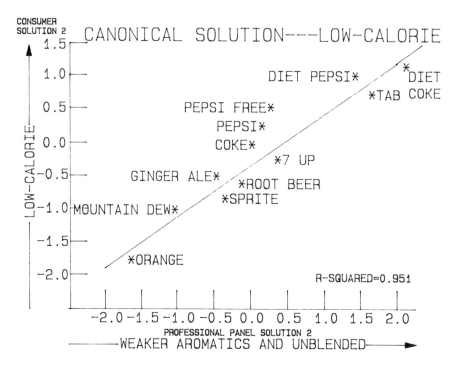

FIGURE 3.   Correlation of PAA data with data from consumer survey of soft drinks, using Canonical Correlation.

In many studies, such as measurement of change in sensory attributes with storage at accelerated temperature, it is frequently impossible to include the control with the samples presented to the panelists. The PAA characterization as summarized on the response sheet provides an excellent and comprehensive summary of the sensory attributes of the baseline product. This goes a long way toward eliminating the problem of "drift" in memory of what the control really tasted like.

D. Statistical Procedures

Many common statistical software programs can be used to analyze PAA data, for example, Principal Components Analysis, ANOVA, and Response-Surface Methodology. Choice of method depends upon the questions asked during the study. For example, it frequently is important to learn how closely a competitor's product compares. Principal Components Analysis can be used for this purpose. One of the interesting outcomes of Principal Components Analysis is perceptual flavor maps. These can be created using the first two or first three components (indices) as coordinates. This technique allows for portrayal of the relative positions of a diverse group of products within a product category, based on the attributes that comprise the indices.

PAA data can be correlated with other data sets such as chemical analyses and consumer test responses. For example, using a Canonical Correlation procedure, data obtained from a professional PAA panel were correlated with data obtained from 300 consumers of soft drinks in Boston in a 1983 study (see Figure 3).[15] The regression index indicated very close agreement between both groups in a ranking of 12 carbonated beverages. The professional panel described the diet colas as having weaker aromatics and being unblended, which was characteristic of all the synthetically sweetened beverages. The consumers were undoubtedly reacting to the same phenomena, but their flavor vocabulary was more limited and they used the summary term "low calorie".

PAA is extremely useful in determining the effect of raw materials and/or formula modification on each flavor attribute, as well as the interaction of effects on the total score of a product.

E. Advantages of PAA

The PAA system has many advantages as compared with Flavor Profile or other descriptive methods of sensory analysis:

1. An experienced Flavor Profile panel can easily be trained to use PAA, requiring only a familiarization with the system and the attribute scales.
2. It provides an efficient means to examine a large number of samples and quickly identify the major perceptual differences.
3. The findings are amenable to several statistical analysis programs.
4. The PAA approach is applicable to many flavor research programs where large numbers of samples must be tasted over an extended period of time.
5. PAA data sets can be correlated with consumer responses, with process or ingredient changes, and with chemical analyses.

PAA is a powerful tool for portraying significant differences in sensory properties among a large group of products. It is based on and can extend Flavor Profile Analysis which is more heavily focused on detailed sensory description of a small set of individual samples.

F. Uses of PAA

PAA is extremely useful in determining the influence of raw materials and/or formula modification (cost reduction) because the effect of the variable on each attribute, as well as the interaction of effects on the total score of a product, are reflected.

The method currently is proprietary and has not been taught to many other practitioners. However, there is increasing interest in applying PAA to snacks, cookies, fruit drinks, and distilled spirits. Correlations of PAA findings with consumer response and chemical analysis have been reported by Arthur D. Little, Inc.[16] All food product categories are amenable for PAA analysis, and much wider application is anticipated for the future.

V. PANEL TRAINING PROGRAMS

Successful application of Flavor Profile Analysis for solving problems in food and related industries has resulted in the continual need for training company personnel in this method. The goal of all training programs is to provide an in-house capability to perform sensory evaluations in a comprehensive, quantitative, and objective manner. Each training program should be designed specifically to meet the needs of a given company, so the scope, content, and duration of each program will vary depending on whether the major objectives are product quality control, new product development, or other research purposes.

The most comprehensive program consists of six phases conducted over a 6-month period:

1. Selection of trainees
2. Introductory course of instruction
3. Advanced course of instruction
4. Monitored work program assignments
5. Post-instructional guidance (follow-up visits)
6. Panel leader recommendations and certification of trainees

After completing this program, the new flavor profile panel will be able to assist in monitoring product quality, guiding new product development, and monitoring competitive products. A yearly review session is very useful to provide the panel with additional guidance and to audit panel performance.

A. Selection of Trainees

1. Criteria

In order to select a group of 8 trainees, as many as 30 candidates should be screened for potential panel membership. It is usually unwise to choose people with administrative or other official responsibilities because they will be unable to attend panel meetings regularly. Those who are chosen should be available on a daily basis and should have reasonable expectation of continued association with the company to justify their training.

Other criteria for panel membership require that the individual have normal acuity and perception, above-average interest in odor and flavor work, and the ability to work cooperatively with others in a group setting. These personal qualities are determined by test scores and through a personal interview.

Selection of trainees is done by a training staff, which administers a series of odor, taste, and perceptual tests and conducts a personal interview with each candidate. The tests provide essential information about the candidates' interest, acuity, perception, and articulateness.

2. Screening Tests

Six screening tests are used to determine the candidates' taste and odor ability and their perceptual style. These tests have been selected for their value in selecting panelists with high potential for sensory evaluation. The materials used and the scoring are not discussed here to avoid influencing future candidates.

Identification test — In this test, the candidate is given a false suggestion in order to determine his or her independence of perceptual judgment.

Basic taste test — This test presents the candidate with the four basic taste solutions (sweet, sour, salty, and bitter), one blank, and one duplicate basic taste. The concentrations are above threshold levels and are equivalent to a slight intensity (numerical symbol 1). A duplicate sample is included to test consistency of response. While the lack of ability to sense these basic taste impressions is rare, a confusion between description of sour and bitter taste is often encountered. This test is administered primarily to determine whether the candidate displays this confusion.

Arrangement test — This test determines the candidate's ability to rank five samples of the same flavor type in a meaningful flavor order and to describe the basis for such ordering. The samples include differences in the integrative aspects of flavor (balance and body or amplitude), as well as the analytical aspects.

Ranking test — This test requires that the candidate perceive different levels of sweet taste in a flavored medium, simulating actual flavor panel performance where panelists have to isolate elements from a complex whole.

Odor recognition series — Twenty odorants are presented to the candidate in 1-oz screw-cap bottles with cotton plugs. The candidate is asked to smell each bottle and define the odor. In ten cases he or she is allowed to select a descriptor from a multiple choice listing, and in ten cases the candidate must supply his or her own descriptor. This combined multiple-choice, open-ended format has two advantages: (1) the inclusion of the multiple-choice items makes the test less frustrating to the candidate, and (2) this procedure distinguishes between candidates who have little ability in recognizing odors, those who can recognize odors but have difficulty in labeling them, and those who have excellent ability to remember and describe a particular sensory impression. The procedure also indicates the candidate's degree of creativity by making him grope for an adequate descriptor.

Perceptual style tests — Two standardized tests used in psychological testing, the Group

Embedded Figures Test and the Myers-Briggs Type Indicator, are given to the candidates. These tests measure the balance of analytic and integrative styles of perception.

In addition, each candidate is rated in five other ways including how the candidate applied himself in taking the tests, the response to the directions given, level of confidence, and interest and attitude toward the tests. Since the tests are given to a group of six candidates at a time, they also provide the opportunity to observe group interaction.

After the candidates have taken the screening tests, they are interviewed about their academic, work, or personal experiences in sensory or associated areas. Other important factors are discussed, such as their interest in performing sensory testing, their time availability, and any allergies to or moral constraints about any food or beverage products. During the discussion the interviewer tries to obtain a sense of the candidates' ability to communicate, as well as their enthusiasm, confidence, imagination, and cooperativeness.

Recommendations based on test performance, the interview impression, and the proctor's assessment are presented to the company, which then can select at least eight panelists. This large number of candidates allows a company to form two panels of four members each or one panel of five or six on a rotating basis.

B. Panel Training Courses

1. Introductory Course

The introductory course consists of lectures and demonstrations concerning the nature of taste and smell; basic requirements for panel work; the techniques, disciplines, and procedures required for reproducible smelling and tasting; and the development of terminology through the use of reference standards.

2. Advanced Course

The advanced course covers additional aspects of sensory evaluation including flavor situations of a more complex nature, reporting of profile findings, and interpretation of results for management use. This course is given approximately 2 months after the first course.

Emphasis is placed on maintaining a relaxed, congenial atmosphere with a good deal of practice and trainee involvement in flavor profiling and feedback from the instructors. From a socio-educational point of view, these principles are consistent with current thinking about how people learn best.[17] Contemporary writers have emphasized the importance of factors such as a nonthreatening environment, student involvement and initiative, experimental learning, practice and feedback, and lectures combined with demonstrations.

Perhaps most important, however, is the fact that the panelist function involves, among other things, the mastery of a new language. As in the study of a foreign language, constant practice and repetition is required before it can be mastered. Constant practice and use is the essential element in the learning of any skill, including flavor profiling.

3. Monitored Work Program Assignments

Trainees improve their ability and build confidence through a series of exercises they must perform daily between the training sessions. The group is required to evaluate different food products in order to practice using the flavor profile and to develop terminology. The products used in early assignments are selected for their proven value as teaching aids. These products frequently include peanut butter and mayonnaise, for example, because of their relatively simple and somewhat familiar flavor construction. Later in the program, the assignments are based on the future activity of the panel. All products used in the training exercises are also profiled by an experienced panel.

The role of panel leader is rotated weekly among the trainees so that all may appreciate the responsibility of, and the cooperation required by, the panel leader. Each student panel leader submits a written report after the assignment has been completed. Observation of each trainee during the courses of instruction and the follow-up visits, as well as evaluation of his or her flavor profile report, provide insight into the trainee's ability in flavor profiling and potential to lead the panel on a permanent basis.

4. Post-Instructional Guidance and Follow-Up Visits

Every 4 to 6 weeks after the advanced course until the end of the 6-month program, the training staff meets with the panel for 1 day to review the principles and techniques of flavor profile and to resolve questions and problems which might have arisen during the period of homework study. Flavor profiles and the reports prepared by the student panel leaders are critiqued and additional products are jointly evaluated. These visits provide the training staff with an opportunity to appraise the growth and development of the group into an effective flavor panel.

5. Panel Leader Recommendations and Certification of Trainees

Recommendation of a permanent panel leader(s) is made at the next-to-last follow-up visit, giving the company an opportunity to establish a working panel before the end of the program. On the final follow-up visit, certificates of achievement are presented to the trainees in a graduation ceremony.

6. Implementation

The training of a flavor profile panel should be looked upon as an investment in a scientific instrument which requires proper use and maintenance. Management's understanding and support of this concept is absolutely necessary if a panel is to function reliably and to continue on a long-term basis. Therefore, it is important that at least one member of management responsible for the flavor panel function attend the training courses as an auditor in order to gain a sound knowledge of the panel's function.

To maintain effectiveness, flavor panels should meet frequently, have defined roles in research projects, and be told the outcome of projects in which they have participated. Also, communciation between the providers of flavor profile information and the users of that information is essential. Review sessions conducted on a semiannual or annual basis refine panelists' performance and reinforce their understanding of the Flavor Profile approach to sensory evaluation. Further, it is frequently helpful for marketing and manufacturing managers to attend a one- or two-day sensory orientation course which summarizes the principles, uses, and limitations of sensory measurement. This course provides managers with a minimum "flavor" vocabulary to use in communicating with sensory project managers. It also provides an insight as to the rigor, objectivity, and accuracy of the Flavor Profile Method.

In conclusion, successful application of Flavor Profile Analysis in developing and improving food products and in solving flavor and odor problems requires panels that can operate efficiently and reliably. This effectiveness of operation is possible only through comprehensive training programs conducted by a staff thoroughly experienced in the Flavor Profile Method.

REFERENCES

1. **Cairncross, S. E. and Sjostrom, L. B.,** Flavor profiles — a new approach to flavor problems, *Food Technol.,* 4(8), 308, 1950.
2. **Caul, J. F.,** The profile method of flavor analysis, in *Advances in Food Research,* Academic Press, New York, 1956.

3. **Caul, J. F., Cairncross, S. E., and Sjostrom, L. B.,** The flavour profile in review, *Perfum. Essent. Oil Rec.,* 49, 130, 1958.

4. **Sjostrom, L. B., Cairncross, S. E., and Caul, J. F.,** Methodology of the flavor profile, *Food Technol.,* 11(9), 20, 1957.

5. **Tajfel, J.,** Social and cultural factors in perception, in *The Handbook of Social Psychology,* Lindzey, G. and Aronson, E., Eds., Vol. 3, Addison-Wesley, Reading, Mass., 1969, 315.

6. **Lewin, K., Lippitt, R., and White, R. K.,** Patterns of aggresive behavior in experimentally created "social climates", *J. Soc. Psychol.,* 10, 271, 1939.

7. **Walsh, D. H.,** Social psychological considerations in flavor measurement, in *Flavor: Its Chemical, Behavioral and Commercial Aspects,* Westview Press, Boulder, Colo., 1978, 144.

8. **Rosenthal, R.,** *Experimenter Effects in Behavioral Research,* Irvington, New York, 1966.

9. **Miller, I.,** Statistical treatment of flavor data, in *Flavor: Its Chemical, Behavioral and Commercial Aspects,* Apt, C. M., Ed., Westview Press, Boulder, Colo., 1978, chap. 11.

10. **Syarief, H., Hamaan, E. D., Giesbrecht, F. G., Young, C. T., and Monroe, R. J.,** Interdependency and underlying dimensions of sensory flavor characteristics of selected foods, J. *Food Sci.,* 50, 631, 1985.

11. **Anon.,** A matter of taste, Investor News & Report, Abbott Laboratories, Abbott Park, Ill., 1984.

12. **Krasner, S. W., McGuire, M. J., and Ferguson, V. B.,** Taste and odors: the flavor profile method, *J. Am. Water Works Assoc.,* p. 34, March 1985.

13. **Bartels, J. H. M., Burlingame, G. A., and Suffet, I. H.,** Flavor profile analysis: taste and odor control of the future, *J. Am. Water Works Assoc.,* p. 50, March 1986.

14. **Caul, J. F.,** The nature of flavor, *Cereal Sci. Today,* 12(7), 273, July 1967.

15. **Hanson, J. E., Kendall, D. A., Smith, N. F., and Hess, A. P.,** The missing link, *Beverage World,* 102, 108, 1983.

16. **Kendall, D. A., Thrun, K. E., and Smith, N. F.,** Product Dimensions — Chemical and Sensory Correlation, paper presented at Institute of Food Technologists Annu. Meet., Anaheim, Calif., June 1984.

17. **Levin, K., Lippitt, R., and White, R. K.,** Patterns of aggressive behavior in experimentally created "social climes", *J. Soc. Psychol.,* 10, 271, 1939.

Chapter 3

QUANTITATIVE DESCRIPTIVE ANALYSIS

Katherine L. Zook and Jacqueline H. Pearce

TABLE OF CONTENTS

I.   Quantitative Descriptive Analysis: What It Is, How the Technique
Was Developed, and Why 44

II.   Uses for QDA® Descriptive Information 47

III. Present Degree and Scope of Use by Industry and Others 47

IV. How to Establish a QDA® Program... 48
    A.   Planning the QDA® 48
    B.   Recruitment and Screening of Panel Members 49
          1.   Recruitment... 49
              a.   In-House Recruitment 49
              b.   Recruitment of Consumers as Judges..................... 49
          2.   Screening Panel Members.. 49
              a.   Setting up the Screening Tests 49
              b.   Conducting the Screening Tests 51
              c.   Selection of Judges After Screening 51
    C.   The Training Sessions and Development of the Scoresheet.............. 51
          1.   Types of Training Products Needed............................. 53
          2.   Setting up a Flexible Schedule for Training..................... 53
          3.   Planning a Single Training Session 53
          4.   The First Session 54
          5.   Later Sessions 54
          6.   Development of the Scoresheet and Definitions................... 55
              a.   Quadrant Training Technique to Obtain Good
                   Panel Participation... 55
              b.   Rank Order Technique to Obtain Good
                   Panel Participation... 55
          7.   Integration of All Terms into One Scoresheet.................... 56
          8.   The Replicated Pilot Test 56
          9.   The Remedial Session or Sessions 56
    D.   The Replicated Product Descriptions 57
          1.   Setting up the QDA® Test 57
              a.   Statistical Design/Randomization Plan..................... 57
              b.   Number of Products, Judges, and Replications............ 57
          2.   Plans for Presentation and Serving of QDA® Products........... 58
    E.   Statistical Analysis of the Descriptive Data........................... 58
          1.   Transferring Data from Scoresheet to Computer 58
          2.   Software and Printout of Data 59
    F.   Interpretation of Results 61
          1.   Product Differences 61
          2.   Judge Perfromance... 62
              a.   Panel Showing Good Performance 63

b.   Panel Showing Poor Performance 63
G.   Presenting a QDA® for Maximum Effectiveness 63

V.   The Use of QDA® Data in Connection with Other Tests Which
Measure Acceptance of the Same Products 66

VI.   The Use of More Complex Statistical Techniques with QDA® 67

VII.   Maintenance of Trained Panels 68
A.   Panels in Continuous Use 68
B.   Panels that are Used on an Intermittent Basis 69

VIII.   Training and Characteristics of a Good QDA® Panel Leader 70

IX.   Relative Strength vs. Difficulties of Using the QDA® Technique 70

References 71

I. QUANTITATIVE DESCRIPTIVE ANALYSIS: WHAT IT IS, HOW THE TECHNIQUE WAS DEVELOPED, AND WHY

QDA®, or Quantitative Descriptive Analysis, was developed in the mid-1970s to answer a need for a technique which could be used describe the sensory characteristics of a product with precision in mathematical terms. With this technique, statistics can be used to measure variability and to compare or contrast one product to other products. The first article about QDA® was published by its developers — Herbert Stone, Joel Sidel, Shirley Oliver, Annette Woolsey, and Richard C. Singleton, then from Stanford University — in 1974 in *Food Technology*.[1] It described the rationale behind the technique as well as its main features. The developers, who worked with the technique for about 5 years before publishing, paid particular attention to the psychological soundess of the scale used to obtain the information from the panelist.[2,3] The technique has since been updated by the originators;[4,5] the updated version will be described here.

Two features of the QDA® technique have set it apart from earlier descriptive techniques such as the Flavor Profile.[6,7] These are the unstructured scale and the spider web method for displaying the data.[1]

The unstructured scale used in QDA® is a 6-in. line anchored only at the ends by marks $1/_2$ in. from either end and by terms which define and limit the attributes. The task of the judge is simply to make a mark across the line at the point which represents what he perceives. Numerical equivalents are assigned later in the process of coding or entering data into the computer (Figure 1).

The other feature which has become associated in most people's minds with QDA® is the spider web method of displaying the data. This is a graph with lines representing the sensory attributes radiating outward from 0 at the center to 60 (or some other value) at the edge. The numerical mean ratings for each attribute are located at the proper points on these

NAME _____   DATE_____   CODE_____

FLAVOR

PLEASE PLACE a vertical line across the horizontal line at the point which best describes
that attribute in the sample.

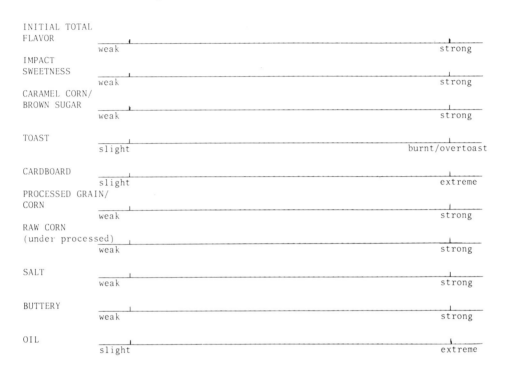

INITIAL TOTAL
FLAVOR
weak   strong
IMPACT
SWEETNESS
weak   strong
CARAMEL CORN/
BROWN SUGAR
weak   strong

TOAST
slight   burnt/overtoast

CARDBOARD
slight   extreme
PROCESSED GRAIN/
CORN
weak   strong

RAW CORN
(under processed)
weak   strong

SALT
weak   strong

BUTTERY
weak   strong

OIL
slight   extreme

FIGURE 1.   Scoresheet for Quantitative Descriptive Analysis. The judge evaluates the intensity of each sensory attribute by placing a vertical line across the unstructured line. This intensity is later converted into a numerical score ranging from 0 to 60 by use of a template or digitizer.

spokes and are connected by lines to give a graphic picture of the descriptive results. This presentation makes it possible to see at a glance the differences and similarities of several products displayed on the same spider web (Figure 2).

Since these two features set QDA® apart from earlier techniques, such as Flavor Profile, many people assume that any descriptive work done with an unstructured scale, analyzed by analysis of variance, and displayed on a spider web is a QDA®. In fact, there is a whole system of procedures and statistical analysis which characterizes the classic QDA® method and ensures its precision and reliability.

By definition, then, QDA® is a technique in which trained individuals identify and quantify the sensory properties of a product or ingredient in order of occurrence.

To summarize, the essential factors in a QDA® are

The use of the unstructured 6-in. scale — The judges mark the intensity of the product characteristic on the scale.

The use of judges screened and trained on the specific product to be described — Screening is by some form of discrimination test, either a triangle or duo trio. Training of the best performers selected for the sensory panel is for 1 hr each day for whatever time is necessary, usually five to seven sessions.

The use of a descriptive score card developed by the panel after seeing many reference

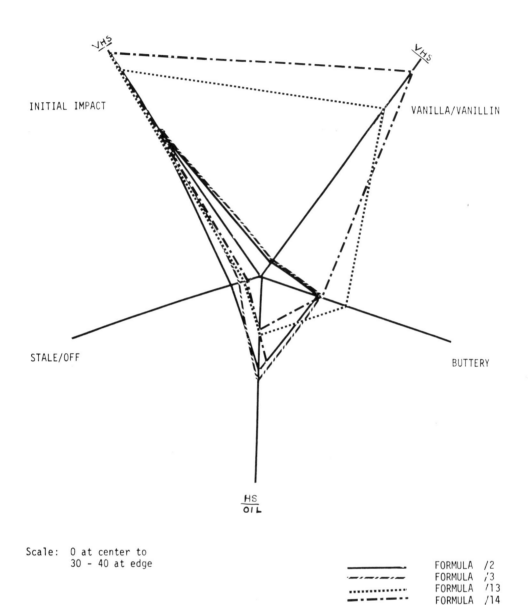

FIGURE 2. A typical QDA® configuration for aroma of a sample. The average intensities for the various attributes are graphed on lines radiating outward from a value of 0 at the center point to a value of 30 at the outward perimeter. Four samples are displayed on the graph, two of which are significantly different than the others.

products and product variations — The score card and judges are evaluated on a pilot test of known products before proceeding with the real test.

The use of a trained panel leader

The use of repeated evaluations of the product by judges seated in separate booths — There is no group pressure to come to a consensus or group judgement.

The use of statistical analysis by analysis of variance which gives average ratings and standard deviations for each attribute and uses tests of significance to distinguish

between products — Information is available on the variability of the panelists and the degree of correlation between the attributes used to describe the products. It is possible to use many sophisticated techniques with QDA® data if special designs are set up at the outset.

The use of the spider web method of graphically displaying the data — This is usually backed up by supporting tables of the actual product means for each attribute arranged in descending order with associated tests of significance. QDA® data can be displayed in many other ways, but this particular method has proven to be an easy-to-understand format for presentation of the sensory data to groups as varied as research, quality control, and marketing.

Since QDA® was first introduced, it has won adherents who like it because it is product-specific, very precise, and adaptable to a whole range of uses and statistical analysis. It also has had detractors who feel that it is too time-consuming and costly, preferring techniques based on industry standards or the earlier Flavor Profile method developed in the 1940s and 1950s.

Having been introduced to the technique by its developers and having worked exclusively with it for some 7 to 8 years,[8] the purpose of the author and co-author in writing this chapter is to explain how to do a classic QDA® and to clarify many peripheral points about the technique.

II. USES FOR QDA® DESCRIPTIVE INFORMATION

Originally, QDA® was considered to be primarily a tool to assist research and development in describing new products or changes in formulations, primarily in food and beverage companies. As it became apparent that *any* sensory attribute could be described — the feel of a fabric, the coolness of a mouthwash — QDA® usage has expanded greatly. It also became apparent that QDA® provided a way for different parts of the company to look at products and exchange information.[4]

Research and development — QDA® provided a way to describe new products, changes in formulation, and methods of manufacture. It became an especially useful tool when ingredients became unavailable or a product was to be cost-reduced.[8] It was also used to check the effect of storage over time, e.g., to determine how fast a product became stale.

Quality control — periodic checks using QDA® helped to determine if the company was still making the same product and to make interplant comparisons. Here, a range or band of product showing the permitted limits of variability proved helpful. These uses helped to minimize "drift" due to operator, machinery, or ingredients. QDA® information could also describe the effect of changing processes or speeding up lines.

Marketing — corporate marketing departments used QDA® to describe competitive products, monitor ongoing changes in them, and document claims (e.g., our syrup is twice as thick as the competitor). Where market share was decreasing, they used it to determine if anything was wrong with the product before going on to other factors.[8] In connection with consumers' perceptions and comments, it could be used to increase certain attributes the consumer noticed or wanted, even to provide terms for advertising.

III. PRESENT DEGREE AND SCOPE OF USE BY INDUSTRY AND OTHERS

Prior to completing this chapter on QDA®, the authors surveyed a sample of 20 major food, beverage, tobacco, and personal care product companies to see if they used descriptive testing in their sensory work. The results indicated that four of the companies used QDA® in its classic sense, following all of the procedures and using the specially developed statistical package for analysis. Another six were using a modified QDA® with their own versions of the screening, training, number of replications and analysis. In some cases, the modifications were efforts to abbreviate the process and make it quicker to carry out. Thus, fully half of the companies surveyed had been influenced by the QDA® method of describing a product quantitatively.

Seven of the companies were using modifications of the earlier Flavor Profile Method, many of whom had adapted the traditional scale (from threshold to slight, moderate, and extreme) so that it was calibrated with numbers and could be analyzed statistically. Of the remaining three, one was using magnitude estimation, one the master brewing system, and one the Odor Profiling Method for describing smoke (American Society for Testing and Materials, Philadelphia).

IV. HOW TO ESTABLISH A QDA® PROGRAM

The first step in establishing a successful QDA® program is to get management support for the idea. Management must be made aware, by conferences, letters, presentations, and individual meetings, how it will effectively promote the firm's short- and long-term goals, what the relative cost and time committments will be, and what is involved in the operational scope of the technique.

A. Planning the QDA®

Once QDA® is in place as a technique and particular project has been decided upon, it is necessary to meet with the requester of the information to define and discuss:

What is the objective of the QDA®? — What questions does the requester have in mind? Is the study being conducted in connection with another type of test, such as a consumer or in-house acceptance test? Understanding the objective will allow for a more efficient test protocol.

Exactly what products are to be included in the QDA®? — Will the array have samples that are similar to each other, or is the test composed of competitive (dissimilar) products? This factor will influence the type of training.

The product quantity needed for the QDA® — Quantities of sample must be calculated for screening, the pretest, training sessions, and the final replicated testings with some left over to review at the presentation of the final results. These requirements need to be reviewed early to make sure that the requester can fulfill the needs.

The availability of screening and training products — Samples that represent the range of product characteristics and deviations from the product provide prospective judges and the trainer with physical illustrations of the issues that may arise when testing begins. The better the quality of the screening and training samples, the more opportunity the panel will have to develop a useful questionnaire and resource set.

A time and responsibility schedule — A definite time and responsibility schedule is necessary to specify when deliveries of products will be made, who on the staff is responsible for given segments, when certain parts of the QDA® will be completed, and when the final presentation will be given.

The QDA® panel leader should be the primary contact in these sessions. The initial discussions allow the panel leader to become aware of the requester's concerns and allow an exchange regarding product issues.

The various steps in conducting a QDA® will now be discussed in the order in which they are carried out — the recruitment and screening of the panel members,[8] the training sessions and development of the scoresheet, the replicated product descriptions, the statistical analysis of the descriptive data, the interpretation of the results, and finally the presentation of the QDA® for maximum effectiveness. The examples of QDA®-evaluated products in the text will be restricted to food products to allow for consistency. QDA® intensity measurements can be adapted for any other stimuli.

B. Recruitment and Screening of Panel Members

1. Recruitment

Panelists for QDA® can be recruited either from the in-house personnel of a company or by obtaining consumers through telephone recruitment or the use of an outside agency.

It is usually necessary to recruit two or three times as many people as you wish to train, both to get the level of discrimination desired in the triangle test and to provide for those who, for unforeseen reasons, drop out or cannot serve. If you wish to train 12 people, it will be necessary to screen about 30 to 36 people.

Whether you are obtaining candidates from within or outside the company, the desired end result is a group of conscientious, perceptive panelists who normally would like or use the product type.

a. In-House Recruitment

The primary concern with internal panels is the judges' time away from the job, particularly for the numerous 1-hr training sessions. If the technique is new to the company, it may be necessary to provide some information about what the technique is and how it functions to improve understanding of the time requirements. Both the person you want as a panelist and his immediate superior must be convinced that this is a worthwhile, company-desired activity.

Research, technical, and business personnel all make good panelists. Middle- to upper-management personnel can be good panelists but are usually unavailable for the time involved. Plant production personnel are usually not good candidates because their absence can hold up the line. It is not a good idea to recruit too heavily from one department, since if all qualify, the area may be short-handed during training. Candidates who agree to participate will need a schedule of dates and times for the various activities involved.

b. Recruitment of Consumers as Judges

The primary concern with consumers is getting the prospective candidate, if he or she qualifies, to the test site time after time. The test site must be central, the candidate must have transportation, the times of testing must be set when the candidates can get there (evenings and weekends included), and the remuneration must be high enough to get a person to return for the whole procedure if he passes the screening tests. Outside judges are usually paid somewhere between 10 and $25/hr for their time, depending on the going rates in the area.

Usually, the panelists recruited are qualified by age and as users of the product to make sure they have purchased it within a certain time period. Telephone recruiting is followed up with a confirming letter for each period of use of the panelist.

If outside judges are brought into a company sensory laboratory, some thought must be given as to how they will be handled to preserve the confidentiality of company information.

2. Screening Panel Members

This section on the screening process for QDA® will focus first on the process of setting up the screening tests, then on how to conduct the tests, and finally on how to select the judges following the testing.

a. Setting up the Screening Tests

The purpose of screening is to select the candidates who are most perceptive to differences in the specific product being described. This can usually be done if 18 to 20 discrimination tests are administered to each prospective judge. Since each test is administered twice, this would require nine to ten pairs of samples with recognizable differences.

Two types of discrimination tests are commonly used, either of which can be selected for the screening:

Table 1
RANDOMIZATION
PLAN FOR
TRIANGLE TEST
USED IN
SCREENING FOR
QDA®

Judge[a]	Set 1	Set 2
1	AAB	BBA
2	BBA	ABA
3	BAA	BAB
4	ABA	ABB
5	ABB	BAA
6	BAB	AAB
7	ABB	BAA
8	ABA	ABB
9	BAB	AAB
10	BBA	ABA
11	AAB	BBA
12	BAA	BAB

[a] Each judge completes two sets of tests, receiving an A sample as the odd sample one time and the B sample as the odd sample one time. Continue in blocks of six for 36 judges.

The triangle test[8] — In this test, the panelist is presented with three coded samples, two of which are alike and one of which is different. The panelist is asked to select the different sample. Individuals usually repeat the tests twice in one sitting, one with two A samples and one with two B samples. The order of serving is carefully controlled as shown in Table 1.

The duo trio test with balanced replication — For each test the subject is provided with an identified reference (R) and two coded samples. The subject selects which of the two coded samples is most similar or matches the reference. For the second presentation, the reference sample may be rotated with the sample not used as the reference used as (R). Again the test is controlled with a carefully balanced plan of the order of serving.

In looking for products for the screening tests, it is valuable to have a good target or reference product and a broad range of products with characteristics covering most of the major differences to be described in the final QDA®.

For example, screening for a ready-to-eat cereal QDA® should include not only flavor differences, but appearance and texture differences as well. The best rule in balancing the screening samples is to make sure that the sensory area which is crucial to answering the test objective is covered thoroughly. Thus, if crispness is one of the attributes which is vital to the test issue, several of the discrimination tests should concern themselves with crispness. Sometimes it is necessary to have products manufactured for this purpose. Other times products can be obtained at retail from one of the manufacturing plants or from what researchers have on hand.

After the entire group of tests is set up, they should be reviewed by the sensory analyst who should rank them in the order of difficulty. The first test is usually an intentionally easy task. This is not only to assess the level of perception of the panelists, but also to give them the experience of a relatively successful first test to keep them from becoming discouraged. Although the tests will still screen the candidates regardless of their order, the panel leader can learn much about the perception of future panelists by watching the results of the ordered screening tests. Panelists also tend to learn about the product as successively harder tests are administered.

A continuing record of test results showing the number of correct pairings for each individual and the panel as a whole will show how difficult the panelists are finding the tests (see Table 2).

b. Conducting the Screening Tests

The screening tests should be conducted in separate booths, each provided with adequate lighting, ventilation, and freedom from odors, noise, and interference, as is basic to all sensory testing.[9] If consumers are to be tested in a location not primarily used for sensory testing, rectangular tables can be set up facing the walls and temporary partitions made of wallboard or some type of expanded stiff styrofoam-like material (1 in. thick) can be put in place. Partitions do not have to extend to the ceiling but should extend beyond the table's edge so that panelists cannot see or talk to one another. All preparation is done in an adjacent room out of sight of the panelists.

Panelists will be provided with pencils, scoresheets, water, and unsalted crackers to clean the palate before, during, and after the grading. Stop watches may be necessary.

A typical order of presentation for a triangle test is given in Table 1. Samples can be identified with three-digit codes which can be imprinted or marked on small self-stick labels applied directly to the serving dishes.

For in-house panelists who come once per day, the three samples of the first triangle test can be given and, after a short interval, the second set or repeat can be served. Where outside consumer judges are being paid by the hour, the first group may be served, then the judges may wait in the waiting room for 10 min before the second group is served and again before the third group is served. This makes it possible to do as many as three pairs of triangle tests during one day's visit.

c. Selection of Judges After Screening

The top 12 to 14 panelists are selected by percentage of correct responses in the screening tests, hopefully taking those with records of 75% correct or better. The results will depend on the difficulty of the tests and also somewhat on the composition of the product. If the product is very simple, for example grits, differences may be easy to see. If the product is more complex, such as pizza, it may be much harder to find the difference being tested and pair correctly. As a result, it may occasionally be necessary to take someone with a record of 66 to 67% correct responses, or even slightly lower.

In addition to performance, availability is very important in judge selection, because the panelists must be able to come to all of the sessions. In making the final selection, the panel leader should consider group dynamics or how the group will work together. If the choice is between two equally perceptive panelists, it is advisable to select the one with the better verbal skills. Other characteristics of good descriptive panelists, such as good health and ability to deal analytically with the text situation, are discussed in Reference 8.

Panelists should be given a letter of notification which will specify the exact times and places of the training sessions and the purpose of the testing. Those who were not selected should be given a letter of thanks with the suggestion that they might be called again for another product testing for which they would quality.

C. The Training Sessions and Development of the Scoresheet

The training of the panel members, who collectively describe the product, and the de-

Table 2

RESULTS OF SCREENING BY TRIANGLE DIFFERENCE TEST USING VARIATIONS OF A DRY CEREAL WHEN TESTED USING 2A SAMPLES AND 2B SAMPLES

Typical Plan for Screening by Triangle Difference Test

Judge	Day 1 Control vs. low sweetness with		Day 2 Control vs. high sweetness		Day 3 Control vs. no salt		Day 4 Control vs. no oil		Day 5 Normal vs. underdried		Day 6 Normal vs. overdried		Day 7 Control vs. high salt	
	2As	2Bs	2As	2Bs	2As	2Bs	2As	2Bs	2As	2Bs	2As	2Bs	2As	2Bs
1				X[a]	X	X	X		X		X		X	
2						X		X			X	X		X
3				X		X	X	X			X	X	X	
4			X								X		X	
5	X			X	X		X		X		X		X	
6			X	X				X				X		X
7	X		X	X		X		X				X		X
8							X	X			X	X	X	
9	X		X	X	X		X				X		X	X
10		X		X		X		X		X	X	X	X	
11		X	X		X			X	X		X		X	
12						X	X	X				X	X	X
13		X				X	X			X				X
14					X		X		X		X	X	X	
15			X	X	X							X		
16	X		X	X	X	X	X		X		X	X	X	X
17			X	X		X	X	X	X		X	X		X
18	X		X	X			X					X	X	X
19		X	X			X		X				X		
20	X	X		X	X	X						X		
21	X												X	X
Total incorrect	7	5	10	12	8	11	11	10	6	2	12	14	13	10

[a] Indicates incorrect identifications for each of 21 judges for 7 triangle tests administered. The number of incorrect identifications for the entire panel both when tested with 2A samples and when tested with 2B samples as part of the triangle test is shown in the bottom line.

Table 3
TIME SCHEDULE FOR TRAINING QDA PANEL

Day(s)	Activities
1	Orientation; overall description of product; judges look at a typical product in its entirety, generating term or attributes which describe the product
2—3	Flavor variations and development of flavor portion of score sheet
4—5	Texture variations and development of texture portion of score sheet
6	Appearance variations and development of appearance portion of score sheet
7	Use of complete score sheet containing all terms for appearance, flavor, and texture; elimination of duplicate terms; final decisions on order and definition
8	Use of score sheet with important or test products to make sure it really works
9—10	Pilot study in booths — judges complete two replicated sets of data (one per day) with products served just as they will be in the real test

Note: This sample training schedule is planned to take the better part of 2 weeks.

velopment of the scoresheet containing the descriptive terms is a very important part of QDA®. The role of the panel leader is a major one, although the panel leader does not take part in the descriptions of the products. The panel leader keeps the group functioning, provides standards and training samples as needed, prepares trial scoresheets from the terms, thinks of ways to clarify confusion, and tests and monitors the judges' performance.

Ideally, the panel leader should know the product well, and should have tasted a broad range of samples so that he or she can select the most effective training samples. Other characteristics of a good panel leader are discussed in Section VIII.

1. Types of Training Products Needed

It is helpful in planning for a QDA® to know that four categories of training products will be needed in addition to the test products which are being described (the list below would be appropriate for a test of similar samples within a product type):

1. A good standard product (as an anchor point)
2. Variations with known amounts of changes (to assess whether panelists can perceive intensity differences)
3. Reference samples of specific characteristics or notes (to provide an external reference)
4. Competitive products (illustrating extreme variations in the product category being evaluated)

It cannot be stressed too strongly that it is almost impossible to do a good job of training without having products available which illustrate variations in the various attributes of the product being described.

2. Setting up a Flexible Schedule for Training

The leader must have a general plan for the training which will be set up before the training commences (see Table 3 for one such example). Plans for each day's sessions must remain somewhat flexible and are finalized after reviewing the previous day's results. Experience will show how much can be accomplished in the 1-hr session usually allotted per day.

3. Planning a Single Training Session

The leader must have a plan for each day's training session which includes the following information:

Table 4

**SAMPLE PLANNING SHEET FOR A SINGLE 1-HR QDA® TRAINING SESSION
(1 HR ON DAY 3)**

1. Serve samples with three levels of sugar, all with the same levels of spice (Code A, B, C).
     Score Sheet: eight terms relating to sweetness and spiciness. Use three different colored pencils.
     Objective: Can panelists perceive the differences in sweetness?
2. Serve samples of the cereal with three levels of sweetness, but with variations in the spice level (Code D, E, F).
     Score sheet: same eight terms.
     Objective: Can panelists perceive the difference in spice?
3. Serve test products, control, and two variations (Code G, H, I).
     Score sheet: full flavor score sheet including initial impact and aftertaste terms.
     Objective: How sweet and how spicy do the panelists perceive the test products to be? How different are they from the control?
4. Have available: reference material for each of the spices in the product: cinnamon, ginger, nutmeg.

1.   Which samples will be served and how they will be coded
2.   What questions will be asked about the samples
3.   What trial scoresheets will be used
4.   What references may be needed

The sample plan in Table 4 gives an example of a session planned in training a panel to describe a cereal with variations in sweetness and spice.

At the conclusion of the training session, the leader will work with the judges' individual scoresheets or results to determine whether the panelists are seeing the *real* differences among these products. This point assumes the training samples were well selected and the panel leader *knows* the differences that should have been measured. If the panel leader had anticipated a certain response and does not find it, they may suggest that panel confusion exists. This information will be used in planning the next training session. Some computer programs are now available which may make it possible to have simple analyses of variance performed on the panelists' data before planning the next day's agenda. The analysis would provide a statistical indication regarding the observed reactions. Ranks and mean values can also be used to understand the judges' responses.

4. The First Session

The first session is important for establishing the routine which will be followed in later sessions. Some orientation is needed so that panel members will be aware of the tasks they are to do and who their fellow panel members are. Usually the first day's work is with the standard product. Judges are asked to write down terms which will describe it in its entirety. Attributes for aroma (if pertinent), appearance, flavor, and texture are noted in the order they are perceived.

These terms can be listed on a chalkboard or large paper pad on an easel in front of the panelists, grouping together the terms which are synonyms for the same attribute. Thus, *soggy*, *wet*, *damp*, and *moist* might all refer to the same attribute — the degree of moistness of the product. Verbal input is obtained from *all* of the judges. The terms provided by the judges will be used in making up the first trial scoresheet.

5. Later Sessions

Later sessions focus on appearance, flavor, and texture in detail. After first letting the training group see the product in its entirety, it may be advisable to concentrate on appearance for 1 day, flavor for 1 or 2 days, and texture for 1 or 2 days. In each day's sessions, panelists will be tasting and looking at variations, backed up by references which will help them to

Sample of results obtained on Sweetness of two cereals during QDA training. The panel leader places a mark on the line showing where each of 12 judges have scored two samples varying in Sweetness.

FIGURE 3. Sample of results obtained on sweetness of two cereals during QDA training. The training exercise is displayed on a chalkboard in front of the group.

define and understand the attributes they are seeing. The variations supplied are the ones that have been reviewed and selected by the panel leader. Depending on how the group evolves and the issues raised, other training samples not originally planned for may need to be introduced.

6. Development of the Scoresheet and Definitions

About the 2nd day, panelists can start to use small portions of the trial scoresheet, which contains six to eight terms that relate to one aspect of the product. Panelists may use separate score sheets for each product described or, using colored pencils, they can grade two or three on the same sheet.

As the panelist is working with the various attribute terms, the definitions for the terms are developed. The panel leader may take the ideas of the group and express them in a clear, concise statement, but should always check back with the group to see if that is what they really mean by the term. Standard definitions for commonly used terms in sensory tests are helpful so that the ideas of the panel will fit into generally accepted terminology.

a. Quadrant Rating Technique to Obtain Good Panel Participation

There are several techniques which work well in holding panel interest and in revealing if all panelists are perceiving the product in a like manner. One way is to ask each panelist if he perceives the product to be in the first, second, third, or fourth quarter of the unstructured line. These placements can be displayed on a line on the chalkboard and will indicate to both panel leader and panelists the degree of similarity in perception which the panel has attained at any given time. This technique also reveals the panelist who is seeing things very differently from the group as a whole.

Because of individual variations in taste, panelists will not use exactly the same place on the line to describe a given term. For instance, judges may rate their perceived degree of sweetness on different parts of the line scale. If, however, good training samples are presented enough times, judges will come to rate the products in the same order, placing the less sweet products lower than the moderately and very sweet products. An example of how 12 panelists' judgments might sort out is shown in Figure 3.

b. Rank Order Technique to Obtain Good Participation

A second technique is that of ranking the strength of the attribute. For instance, a panel can grade three samples of sweetness on one scoresheet using colored pencils. They can then give their rankings for the leader to display on the chalkboard as follows:

Objective: Which is the sweetest product?

Judges		1 2 3 4 5 etc.	
Product	1	1 2 1 3 1	Sweetest
	2	3 3 2 2 3	
	3	2 1 3 1 2	Least sweet

It can be quickly established that judges 2 and 4 do not see product 1 as the sweetest product. Discussion can take place between those with opposing points of view and more tasting can take place if necessary to clarify the issue. It is important that the panelists actively work at tasting and describing the product rather than sit passively and listen to others describe what they taste.

7. Integration of All Terms into One Scoresheet

When the panel leader feels that the appearance, flavor, and texture of the product have all been adequately described, he or she will combine the various short sections into one complete scoresheet to be tried out with the test products (about day 7 or 8 of the training schedule).

In preparing the final scoresheet, the leader must be aware of several things:

1. Each term should describe one sensory attribute and be a single continuum. No scales should change from *amount* on the left to an *affective opinion* on the right. For example, instead of *very slightly oily* to *too oily*, the scale should be *slight* (slightly oily) or *low* on the left to *extreme* (extremely oily) or *high* on the right.
2. Redundancy should be eliminated. If two terms describe the same thing, the panel should decide which term to eliminate.
3. The attributes should be graded in a comfortable order, generally the order perceived, for both texture and flavor. The initial crunch of a cereal might come before the pastiness, which develops with chewing. The flavor scoresheet should start with the initial impact followed by the middle flavor terms and end with the aftertaste terms.
4. All panelists must grade the product under the same conditions. Thus, the scoresheet may need special instructions such as "wait 30 sec after adding milk". Sometimes stop watches in the individual booths during testing are required.
5. Each attribute should mean the same thing to all judges. If the term means different things to different judges it should not be used, for the results will be unclear and confused. The panel leader should then try to use another term which means the same thing to all the judges.

At about this time in the training, the definitions should be ready to be finalized. Figure 1 presents an example of a typical QDA® scoresheet.

8. The Replicated Pilot Test

The replicated pilot test is conducted on the last 2 days of the training period. Since it is a trial run for the way the QDA® will actually be conducted, refer to the section on the Replicated QDA® Product Description (Section IV.D) for directions on how to conduct the test.

9. The Remedial Session or Sessions

After statistical analysis of data from the pilot tests, the QDA® leader will review the results for the panel as a whole and for each individual judge. During the last session, sets of samples will be presented which will hlep to clarify any confusion and improve the preciseness of the description. The panel leader must make the decision as to when the panel and the scoresheet are ready for the actual test. This decision will be based on how the panel is meeting the expectations of the leader for description and discrimination of products and how the panelists' attitudes toward performance appear.

D. The Replicated Product Descriptions

The actual QDA® description of the test products is carried out in sensory booths so that each judge's evaluation is private and individual. The evaluation is repeated on several consecutive days, the number of replications depending upon factors such as the number or type of products being described, the number of available trained judges, and the degree of precision desired in the final results. Four replications are generally used with less than 12 judges, while three replications will be sufficient if 12 or more judges are utilized.

1. Setting up the QDA® Test

A number of basic decisions have to be made in setting up the test. These involve determining:

1. What statistical design of randomization plan will be used
2. What products will be included in the description
3. How the products will be served, with what additives, and at what intervals
4. How many judges will be used
5. How many times the data will be replicated

These factors influence the scheduling of panelists and physical presentation of the product. Such decisions are facilitated through the early reviews of the project objectives with the requester and in careful observation during the training sessions.

a. Statistical Design/Randomization Plan

The statistical design will determine the order of serving the product to the individual judge, as well as how the data can be analyzed. To avoid position bias in a two-sample test, each product should be described an equal number of times in first position and in second position. Possible orders of serving for three- and four-sample tests are given in statistical texts.[10]

For more complicated designs where all of the samples cannot be viewed by the judge at one sitting, one replication may take several days to complete, serving as many as four to six samples per day. Special designs must be followed in these cases which are completely balanced for judge, order, sample, and replication. In general, it is better to complete the entire block before repeating any of it in order to make sure that all samples are tested in all positions both at the beginning and end of the testing, thus minimizing time and sequence effects.

b. Number of Products, Judges, and Replications

The number of products to be served at one sitting depends largely on the degree of fatigue or carryover which the judge will experience, as well as the length or complexity of the score sheet. Where the test is not complicated, judges can test four to six samples at one sitting, while for products with complicated characteristics (hot salsa, for example), three products may be the maximum.

Good QDA® information can be obtained with 6 to 8 judges, although 10 to 12 judges seems to be a commonly used number. If too few judges are used, the variability of the test may be so great that large differences between samples will be required to show statistical significance between samples. If too many judges (and replications) are used, the test results may show statistical differences among samples which are of no practical value to the user of the information.

2. Plans for Presentation and Serving of QDA® Products

QDA® samples are usually served in disposable dishes coded with three-digit random number codes. The codes may be preprinted on small self-stick labels and affixed to the dish, or a wax-type pencil may be used to write the codes on the top or side of the serving dish. Polyethylene dishes are usually used because of their disposability and good heat retention, but glass and ceramic are also usable if odor-free and white or neutral in color.

The samples are usually served under controlled conditions similar to normal consumption patterns for the food. This means that temperature of serving must be specified. Electric holding trays or a water bath may be needed for products consumed when warm. Some foods (for instance, pizza) may have to be baked to be ready at specific times when the judges should be there. In these cases, judges must be notified to arrive on time. Additives such as milk on cereal or syrup on pancakes are usually measured. Products may be described both without and with the additive if such information is needed. The precise method and conditions of serving are usually worked out during the training session and tried out during the pilot test.

For most samples, the interval between finishing one sample and being served the next is sufficient, but for highly spiced samples or for those with much carry-over, a stop watch may be placed in each booth and panelists may be required to wait 1 to 3 min. This time interval appears to allow for satisfactory additional testing.

Tasteless, odorless water, either cool or warm, and unsalted crackers or breadsticks are used to clear the palate between samples. Sliced, peeled apples or very mildly flavored cheese have also been used when appropriate.

One or more technicians or sensory persons will be required to physically serve the products to the QDA® judges. The number of judges scheduled to be served at one time depends upon the number of booths available and the amount of technician help. Judges should be scheduled at a rate that allows the technician time to serve them and clean up the booth even if they are delayed 2 to 3 min. The judge should never feel hurried in his evaluation.

Each judge is served according to a predetermined order of testing. If he misses a day, makeup time is scheduled. If this is impossible, the judge is served his samples in consecutive order and will be missing a part of the last replication. Judges must complete at least two replications, otherwise there will be no measure of the variability of the judge for the analysis.

Immediately following an individual test, data sheets must be checked for values which may be missing due to a grading oversight. Generally, the judge can supply the missing values before he leaves the area.

A light snack is often served after a QDA® evaluation in a waiting room or side area which is not a part of the judging area. Sometimes it is effective as a morale booster to plan a special treat or get-together after a particularly long schedule of testing. If security permits, a review of the outcome of the test or how the results affected a decision impacts positively on judge psychology.

E. Statistical Analysis of the Descriptive Data
1. Transferring Data from Scoresheet to Computer

The data are obtained from a QDA® scoresheet by assigning a number from 0 to 60 to the point at which the judge placed his mark across the unstructured scale. One such value is obtained for each attribute measured. Thus, a study of four products (each described on 25 attributes) using ten judges and four replications would yield 4000 pieces of data.

Although it is possible to obtain these numbers by using a measuring device or template, this is very time-consuming. The more usual approach is to enter the data into a computer via a digitizer. The scoresheet is placed on a digitizer, a plastic tablet in which piano wires are embedded both crosswise and lengthwise at about 1/100 in. By touching a stylus to the point on the line marked by the panelist, the numberical value of the coordinate is read

Table 5
ANALYSES OF VARIANCE OF QDA® DATA FOR INDIVIDUAL ATTRIBUTES OF A FOOD PRODUCT

Attributes		Degrees of freedom	Sum of squares	Mean square	Variance vs. error		Variance vs. interaction	
					F-ratio	Probability	F-ratio	Probability
1—Degree of	Wheat	2	16630.691	8315.344	72.660	0.000	23.234	0.000
	Subject	9	3939.890	437.765	3.825	0.000	1.223	0.341
	Interaction	18	6441.988	357.988	3.127	0.000		
	Error	267	30555.797	114.441				
2—Toast flavor	Wheat	2	15579.094	7789.547	234.267	0.000	29.676	0.000
	Subject	9	3990.015	443.335	13.333	0.000	1.589	0.164
	Interaction	18	4724.715	262.484	7.894	0.000		
	Error	269	8944.461	33.251				
3—Crispness	Wheat	2	980.650	490.325	5.642	0.004	7.291	0.005
	Subject	9	5978.621	664.291	7.644	0.000	9.877	0.000
	Interaction	18	1210.581	67.255	0.774	0.730		
	Error	268	23289.746	86.902				

directly into the computer and becomes the score for one judge on one attribute of a single product.

This type of data entry requires several devices: a digitizer with visual display, a storage system such as a disk for holding the information until needed, a microprocessor for information control, a printer to print the data and edit it for accuracy before it is analyzed, and a connection to a main frame computer with appropriate software.

New computer advances are leading to the ability to enter the descriptive data directly to a microcomputer or other receiving device and then transfer the data electronically to a larger or more powerful analysis unit. While much of this approach is in its infancy at the time of this writing, it should provide for faster data transfer and fewer omitted responses.

2. Software and Printout of Data

The software for analysis of QDA® data was developed by the originators of the technique,* through whom it is available, and follows generally accepted statistical methods. The basic analysis is relatively unchanged from the original, but several refinements have been added in the last few years to make it easier for the QDA® leader to evaluate judges' performance.

The output from the QDA® analysis consists of:

1. The data array
2. A table of mean scores for each attribute with standard deviations, F ratios, and probabilities of significance (available for each judge and the panel as a whole)
3. Interaction information including F ratio and probabilities
4. T value differences of panel means for products
5. Rank order by judge and panel
6. Subject performance and probability of finding a significant difference
7. Analysis of variance vs. error and vs. interaction (see Table 5)
8. Table of correlations of panel means — each attribute correlated with every other (see Table 6)
9. Summary table for individual subject ratios and probability values for interaction
10. Table of frequency of mean ranks for each attribute

* Herbert Stone, Joel Sidel, Shirley Oliver, Annette Woolsey, and Richard C. Singleton, Tragon Corporation, 365 Convention Way, Redwood City, Calif. 94063

Table 6
CORRELATION COEFFICIENTS FOR EACH ATTRIBUTE OF A FOOD PRODUCT WITH ALL OTHER ATTRIBUTES

Attribute	1	2	3	4	5	6	7	8	9	10	11	12	13	14	Overall
1	100														
2	99	100													
3	-37	-26	100												
4	-19	-8	98	100											
5	93	88	-69	-54	100										
6	96	92	-61	-45	99	100									
7	-77	-69	88	77	-95	-92	100								
8	14	26	87	94	-24	-14	52	100							
9	-81	-87	-25	-43	-53	-61	24	-70	100						
10	95	98	-7	11	77	84	54	43	-95	100					
11	-100	-98	43	26	-95	-98	81	-7	76	-93	100				
12	-100	-98	42	25	-95	-98	81	-8	77	-94	100	100			
13	100	99	-41	-24	94	97	-80	10	-78	94	-100	-100	100		
14	-100	-100	34	17	-92	-95	75	-17	82	-96	-100	-100	-100	100	
Overall	100	99	-39	-21	94	97	-78	12	-79	95	-100	-100	100	-100	100

Coefficient Variable

Table 7
ILLUSTRATION OF DATA RANK ORDERED IN INTENSITY INDICATING GROUPS OF RELATED VALUES

Sweetness	Spice	Wheat flavor
Product A 23.1	Product A 10.2	Product A 15.1
Product C 15.7	Product B 7.1	Product B 13.2
Product B 14.1	Product C 6.9	Product D 13.0
Product D 10.0	Product D 4.9	Product C 12.5
Highly significant	Significant	Not significant
Product A ≠ Product C, B, D	A = B	A = B = D =
	B = C = D	C
	A ≠ C, D	

Note: The probability of finding a significant difference is indicated below each attribute using Very Highly Significant for 0.001 degrees of probability, Highly Significant for 0.01, Significant for 0.05, and T for significance at a probability of 0.10.

The table of mean scores for the panel as a whole makes it possible to quantify each attribute and to check its variability. The analysis of variance shows what products are significantly different from each other, which are not, and to what degree differences exist.

The performance of the judges is checked in two ways: (1) by the tables giving information on how each individual judge rated each product and if he found significant differences and (2) by the interaction tables which help to distinguish between judges who see the samples differently (crossover) and those who see the same differences but to a different degree.

The table of correlations helps to track relationships between the attributes. For instance, crispness may be positively correlated with crunchiness (as one goes up the other also increases) and negatively correlated with "moist feel" (as crispness increases, moist feel decreases).

F. Interpretation of Results

Once the QDA® analysis has been completed by the computer and has been checked to determine if it is error-free, it is necessary to look at the data in two ways:

1. What does the data reveal about the *products* described?
2. What does the data reveal about the performance of the *judges* and the *variability* between products?

1. Product Differences

One of the first steps is to set up tables of related attributes, listing the products in the order of their numerical ratings for the panel as a whole, placing the product with the highest rating at the top and the one with the lowest rating at the bottom as shown in Table 7. Tests of significance of T values (a statistical value obtained in testing for the difference between two sample means) are carried out using a Duncan Multiple Range Test of Significance (or some other range method). Lines are then drawn connecting the products which are not significantly different and separating those which are. This information reveals in what attributes the products vary in a major way or minor way and in what attributes they do not vary statistically.

A second way to examine product differences would be to set up the typical QDA® spider web graph. A product may require several graphs, one for the appearance, one (or two) for the flavor, and another for the texture. Often, the number and extent of differences between products will show up very quickly when displayed visually as in Figure 2.

With the statistical results and the graphic display of information, the interpreter should look at the data critically asking a number of questions:

1. What were the attributes which varied significantly among the products? Were they highly significantly different, just significantly, or borderline effects?
2. What were the attributes which did *not* vary significantly?
3. What attributes are moving together — both increasing or decreasing as a product change is made?
4. What attributes are moving in opposite directions?
5. Are these results consistent with what is known of the product and the results of earlier tests?

Using these questions is a valuable way to formulate the basic conclusions for the data. It leads to setting up key points in the data that can be expressed as follows:

●   Product A is most spicy but least sweet and least crisp.
●   Product B has about the same degree of spice but is less sweet and more crisp.
●   Product C is less spicy than either but resembles Product B in sweetness and crispness.

2. Judge Performance

Before drawing any definite conclusions from the data, it is necessary to look at other parts of the analysis to determine how the judges are performing. Here the analysis by individual judges is helpful with means and standard deviations for individual attributes, as is the summary table of judge interactions. Some questions to ask are

1. Are there judges with extremely large standard deviations, indicating that they do not grade the product the same way time after time?
2. Do the judges place the products in the same order?
3. Are there some judges who see *no* differences, where the panel as a whole *does* see differences?

Whenever the QDA® program automatically tests the significant effect of the product effects against the interaction instead of against error, there is usually some confusion as to how the judges are grading the product. It is important to determine if the large interaction is due to "crossover effect", where judges *disagree*, or if it is merely a matter of judges seeing relatively small differences in intensity among the samples.[1]

Some judge disagreement is to be expected. The results with the panel as a whole usually override the effect of a single errant judge. It is not usual to "drop" judges in the analysis if they do not appear to be seeing the differences as clearly as others. This puts one in the position of selecting whose data one wishes to use and whose one wishes to discard. However, further training may be arranged for a weak judge, or if his performance is deficient in many areas, he may not be recalled on the next use of the panel.

When the analysis of the QDA® data has been thoroughly examined for the significant product differences and for the judges' performance, the panel's conclusions can be drawn from the results. These conclusions should be set up in terms which will relate back to the objective of the test or the project as a whole.

Two examples are given to illustrate how the information from the statistical print-out is used — one from a panel which would be considered a well-performing panel and one from a poorly performing panel.

a. Panel Showing Good Performance

The print-out shown in Table 8 indicates that for the panel as a whole, Products C and A were both fairly starchy (with intensities of 20.58 and 19.50), but the product B was much less starchy (7.839). All of the panelists followed the same pattern in their scoring, although they used different intensity values to express the relationship of the products to each other. The rankings (the small numbers below the actual mean values) indicate that all of the panelists found Product B to be least starchy.

The test of the product mean ratings against error shows that there was a highly significant difference among the samples. The test of the product/judge interaction against error shows that it was *not* significant, meaning that the judges rated the products in a similar fashion.

Use of the Duncan Multiple Range Test with the t-value differences of the panel means indicates that Products A and C are not significantly different from each other (t-value of 0.90) but that both Products A and C are significantly different from Product B (t-values of 12.03 for C/B comparison and 11.04 for A/B comparison). A standard statistical table for the Duncan Multiple Range Test indicates the values of t necessary for significance at several levels.

b. Panel Showing Poor Performance

The printout shown in Table 9 indicates that for the panel as a whole, Products A and C are moderate in toast intensity (31.28 and 29.08) with Product B somewhat less in intensity (27.561). An examination of the rankings by the individual judges indicates that judges 1, 2, 4, and 8 find Product B to be the *most* intense and that five of the remaining six judges find Product B the *least* intense. This type of disagreement, whether from an individual judge or a number of judges, is known as the crossover type of product/judge interaction.[5]

The results of the analysis of variance further confirm the disagreement among the judges, as there is a highly significant interaction effect (0.000 from Table 9). When this occurs, the program automatically tests the product mean square against the interaction mean square instead of the error mean square.

The Duncan Multiple Range Test indicates no significant differences among the samples (t-value differences of 0.76 between A and C, 1.27 between A and B, and 0.50 between C and B in Table 9).

Review of Table 9 indicates that the test of the produce mean square against error is just below the 95% level of significance (0.045). Had the judges in this study all agreed in their evaluations, it is possible that the variation in sample means would have been found to be significant. In a study where the disagreement is less severe (perhaps confined to one judge) and the differences among products more decided, the test of the product mean square against the interaction may reveal significant differences among the products, even though a significant interaction is present.

G. Presenting a QDA® for Maximum Effectiveness

QDA® results are quite frequently presented both in a visual form with transparencies or slides to the interested group and also in a written report to substantiate the results. Here the spider web graph, which has become the identifying mark of QDA®, has proven to be an effective way of communicating the results to a group, especially if the QDA® contains only two to three samples and is conducted repeatedly on the same product so that viewers become familiar with typical shapes.

A typical spider web graph is shown in Figure 2. It depicts each attribute of the product on a single line radiating outward from the center, which is the 0 or low point of all the lines. All lines become stronger or more intense as they radiate outward from the center and all are considered to consist of 60 points (or some other designated number of points). The numerical values for the mean intensity of each attribute, taken from the statistical

Table 8
MEAN SCORES FOR ONE ATTRIBUTE OF A FOOD PRODUCT — WELL-PERFORMING PANEL

Attribute 14: Mean Scores for Starch

Wheat	Panel	Judges									
		1	2	3	4	5	6	7	8	9	10
Product A											
Score	19.50	28.00	6.90	12.70	17.90	32.20	10.10	15.60	28.40	23.10	20.50
Rank	2	1	2	2	1	1	2	2	2	2	1
Product B											
Score	7.839	11.30	5.20	1.89	0.90	22.60	1.20	10.10	11.40	13.30	0.50
Rank	3	3	3	3	33	3	3	3	3	3	3
Product C											
Score	20.580	26.00	11.40	13.10	14.20	28.60	12.30	16.40	30.50	32.80	20.50
Rank	1	2	1	1	2	2	1	1	1	1	2
Mean	15.986	21.77	7.83	9.23	11.00	27.80	7.87	14.03	23.43	23.07	13.83
Standard Deviation	10.577	9.74	5.29	9.59	16.05	8.11	7.92	5.58	8.28	19.26	6.82
Number in panel	299	30	30	29	30	30	30	30	30	30	30
F-Ratio	44.591	8.76	3.66	4.25	3.10	3.58	5.50	3.77	16.00	2.56	28.63
Probability	0.000	0.00	0.04	0.03	0.06	0.04	0.01	0.04	0.00	0.10	0.00
Interaction											
F-Ratio	1.199	0.63	2.55	0.06	0.84	1.18	0.20	1.32	1.15	2.05	2.02
Probability	0.261	0.43	0.11	0.81	0.36	0.28	0.66	0.25	0.28	0.15	0.16

t-Value Differences of Panel Means (degrees of freedom = 269)

Product C	0.0		
Product A	0.90	0.0	
Product B	12.03	11.04	0.0

Table 9

MEAN SCORES FOR ONE ATTRIBUTE OF A FOOD PRODUCT — POORLY PERFORMING PANEL

Mean Scores for Toast

Wheat	Panel	Judges									
		1	2	3	4	5	6	7	8	9	10
Product A											
Score	31.280	22.40	30.40	47.70	29.40	31.60	30.10	39.50	25.60	22.80	33.30
Rank		2	2	2	3	2	2	1	2	1	1
Product B											
Score	27.566	30.40	45.80	15.30	35.90	25.40	23.60	25.56	28.60	16.30	28.80
Rank		1	1	3	1	3	3	3	1	3	2
Product C											
Score	29.037	14.50	18.90	50.50	34.90	32.00	39.90	34.70	21.30	17.00	26.67
Rank		3	3	1	2	1	1	2	3	2	3
Mean	29.294	22.43	31.70	37.83	33.40	29.67	31.20	33.25	25.17	18.70	29.59
Standard Deviation	10.515	6.30	13.46	8.11	15.36	6.55	11.25	5.29	8.66	16.93	4.21
Number in panel	298	30	30	30	30	30	30	29	30	30	29
F-Ratio	3.144	15.93	10.06	58.23	0.52	3.19	5.32	17.31	1.80	0.44	6.24
Probability	0.045	0.00	0.00	0.00	0.60	0.06	0.01	0.00	0.19	0.65	0.01
Interaction											
F-Ratio	7.729	7.84	20.89	33.04	2.83	0.66	6.20	2.83	2.10	0.35	0.55
Probability	0.00	0.01	0.00	0.00	0.09	0.42	0.01	0.09	0.15	0.55	0.46

t-Value Differences of Panel Mean (degree of freedom = 18)

Product A	0.0		
Product B	0.76	0.0	
Product C	1.27	0.50	0.0

analysis, are plotted on these lines. The values for one product are then connected by a line forming the spider web. For visual presentations, various colors and widths of line-type gummed tape can be used so that each product can be readily distinguished. This type of presentation has many advantages:

The observer can grasp many factors at a glance — For instance, if eight attributes are included in one graph and one product is much crisper than the others, this one significant area of difference will stand out at a glance. Also, if products are nearly identical, this too will show up quickly.

The relative size or intensity of a flavor is depicted — One product can present a much "larger" profile on the graph than another.

In giving visual presentation to a group, products can be graphed on separate transparencies with different identifying lines. These can then be placed over one another and projected onto the screen to give an instant comparison of any two desired products.

The most effective presentations are those which are organized along the objective of the test, relating the results of the test to the objective and to the group to whom it is being presented, such as research or marketing. Because there is so much data, the area of interest to the group may be pinpointed or highlighted and supporting data may be covered briefly.

Graphic data is usually backed up by the numerical data, with the sets of tables of mean ratings for each attribute arranged to show which are significantly different. The actual numbers are very important to number-oriented colleagues. It is sometimes effective to state the results or conclusions first in presenting the data and then substantiate why these conclusions were reached with spider web graphs and mean rating tables.

The spider web type of graph is not the only way QDA® data can be presented effectively. If the QDA® is describing products which have increasing levels of an ingredient intended to make the product more crisp, an ordinary graph showing changes in hardness, crispness, or toughness with increasing amounts of the new ingredient might be effective. An example of representing the data using only text is given in Table 10.

QDA® data can also be presented in a linear type graph with the degree of intensity on the vertical axis and the names of the attributes on the horizontal axis. The intensities for one product are then connected by a line. This is an effective visual display medium when several products have been evaluated or are being reviewed.

V. THE USE OF QDA® DATA IN CONNECTION WITH OTHER TESTS WHICH MEASURE ACCEPTANCE OF THE SAME PRODUCTS

Having acceptance data in combination with QDA® data gives a degree of perspective to the results which neither alone can furnish. For instance, two products may be described in slightly different terms via QDA® and yet may be equally acceptable. Conversely, products may be described as being different in only one or two attributes and this seemingly minor change may make a large difference in how acceptable the product is.

Acceptance data can be obtained:

1.   In an in-house acceptance test using 36 to 50 people who like the product
2.   In a larger consumer test of 250 to 300 qualified consumers of the product

Because of the cost, it is not possible to consumer test every new product or variation of an old product. QDAs® and accompanying in-house acceptance data can be obtained on a number of promising candidates and only the most promising one or two sent out on the larger consumer test.

Some companies use the QDA® description in another way. After the consumer test, they use the comments from the test — information on whether the product is too sweet, too

Table 10
ONE POSSIBLE PRESENTATION
FOR QDA® DATA SUMMARIZING
THE SENSORY CHANGES
OCCURRING WITH A FORMULA
MODIFICATION (FOR VISUAL OR
SLIDE PRESENTATION)

Breakfast Item

Formula Change

Old formula	Modified formula
1.98% Sugar	4.18% Sugar
No vanilla flavor	0.09% Vanilla flavor
No butter flavor	0.02% Butter flavor
4.54% Yellow corn flour	No corn flour

Primary Sensory Differences

Aroma	•	More vanilla aroma
	•	More artificial aroma
	•	More perfumy aroma
Flavor	•	More vanilla/vanillin
	•	More perfumy
	•	More lingering vanilla
	•	With syrup
	•	More vanilla flavor
	•	Less eggy flavor
Texture	•	Moister interior

salty, too spicy, or lacking in flavor — to relate to certain specific attributes on the QDA® to determine if they are moving in the right direction with redevelopment efforts. If parallel QDA® and acceptance testing are planned it is very important that they be conducted with the same batches of products and at the same approximate time so that the products will be of the same age.[11]

VI. THE USE OF MORE COMPLEX STATISTICAL TECHNIQUES WITH QDA®

QDA® data can be used with more complex statistical techniques such as multivariate analysis, factor analysis, and response surface analysis. To do this successfully the designs must be set up ahead of time, and products must be manufactured that will adequately represent the various points of the design.

These designs usually involve description of a much larger number of products which, with the required replications (usually only 2 or 3), cause the QDA® descriptions to require a much longer time than usual to complete. Panelists can usually handle six products at a sitting. If, for instance, the design contains nine samples and three replications were desired, the individual panelist would require 4 to 5 days to complete the sets.

It is desirable for the control or current product to be one of the samples present in the design so that there will be a base product to compare against in interpreting the results. A simple factorial design for a cereal containing three levels of sugar and three levels of toast is shown in Table 11.

Table 11
EXAMPLE OF A 3 × 3 EXPERIMENTAL DESIGN
FOR THE EVALUATION OF SWEETNESS AND
TOAST

Cereal varied in sweetness and toast

Level of sweetness

		1	2	3	
	1				Objective: to describe the effects of increasing
Level of toast	2	Current product			sugar content and surface toast on ready-to-
	3				eat cereal

It is usual to complete the simple QDA® analysis of variance prior to doing the factorial analysis. Thus, information is available both on the individual samples and the overall trends — for example, the relative effect of increasing sweetness and toast on certain attributes of the product alone or in combination with each other. From these descriptions, it is usually possible to select a set of two or three products which would be worth submitting to an in-house acceptance panel or an outside consumer test.

QDA® data can be used with Response Surface designs. However, these designs are complicated and require a sound knowledge of the product and expected product changes in order to be interpreted correctly. They also require some expertise in doing Response Surface Analysis. Again, due to the large number of total samples in the design, adequate planning for the test is required.

Multivariate Analysis can be used on QDA® data, but interpreting the results of such data can be very tricky, as various single factors which show correlations among themselves are combined and designated as significant factors in combination responsible for physical changes in the appearance, flavor, and texture of products.

It is not the intent of this paper to point out in detail how to conduct data with complex designs, but only to indicate that unless the interpreter of the data has a good knowledge of the product and the attributes, erroneous conclusions can be drawn, such as mistaking factors which are *associated together* as causative.

QDA® data can also be used to give information on the same products described a number of times as they age under various storage conditions, for example, to provide information as to how and when staling takes place and what attributes it affects. However, to do this type of work, a well-established, stable panel must be used which remains essentially the same in membership. Each time a test is conducted the panel members will be drawn from the same pool of people. Any new trainees for such a panel have to be carefully monitored before they are added to the group. Graphs can be prepared which will show gradual loss of freshness (or increase in staleness) and other changes such as the development of off-notes (Figure 4).

VII. MAINTENANCE OF TRAINED PANELS

A. Panels in Continuous Use

A panel that is used on a regular basis, participating in a series of tests once a month or more often, is deemed to be in continuous use. As the panel leader examines the QDA® data from the replicated sets of ongoing projects, she also examines the performance of the individual judges. It is thus possible to determine if the panel as a whole is drifting or if some individual judges are losing their sensitivity. Sometimes a continuing record is kept on a few important attributes showing whether the individual judge found the significant

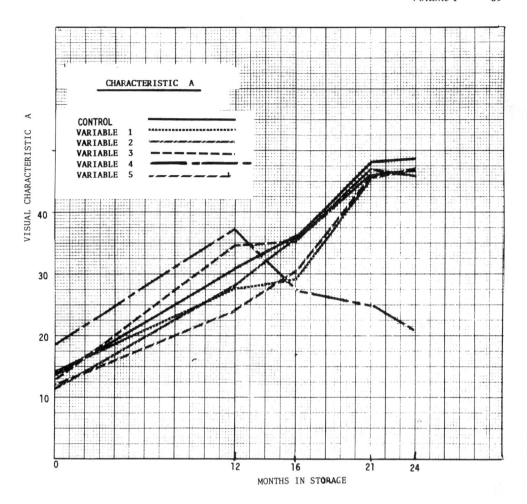

FIGURE 4. Graph of QDA® data on one attribute showing the changes over time for five variables on a 0 to 60 scale.

differences verified by the whole panel. Computerized tracking of the judges' results dramatically assists in this effort.

In many cases there is a control or standard reference product which is or could be one of the products in a planned QDA®. The reference provides the panel leader with good information on how consistently the panel members are grading.

Whenever something new is introduced into a QDA®, such as an off-taste or an experimental product with new attributes, the QDA® panelists may have to be assembled beforehand to become familiar with the new attribute and to add terms to the scoresheet which will describe it. Panelists also should be encouraged to bring up any concerns they may have about products they have been grading or the use of the scoresheet.

B. Panels that are Used on an Intermittent Basis

As sometimes happens, a QDA® panel will be trained and do a considerable amount of testing, but once the initial project is concluded they will not be used for a number of months. When the second project comes up any time after 2 months, the panel will need to be sharpened up again by one or more warm-up sessions, reviewing key products, reference material, and the definitions for the terms. The panel leader should use methods reviewed under the pilot test.

An even better way to keep a panel at top performance is to arrange for them to see and describe some regular production samples about once a month if there are no special projects underway.

Panelists also need occasional feedback from the panel leader as to how they are doing individually or what the panel is accomplishing. Information on current projects should, of course, be kept from the panelists since it could bias the results, but when the work is entirely completed it is sometimes effective to have a session with graphs, charts, or spider web graphs to show some of the information uncovered by the panel.

Whether the panel is used continuously or intermittently, the panel leader must remain alert to the performance of the panel and must motivate them to remain serious and caring rather than to think of the QDA® evaluation as a routine task to be accomplished as quickly as possible.

VIII. TRAINING AND CHARACTERISTICS OF A GOOD QDA® PANEL LEADER

A good panel leader must have training and skills in three areas:

1. A good sensory background and knowledge of food, beverage, or whatever the product is to be described
2. Some knowledge or background in psychology and the practical aspects of group dynamics
3. Some basic knowledge of statistics and the theory of probability in order to be able to interpret the data

The panel leader's purpose is to be an expediter, not a monitor. He or she should be willing to search for and bring in whatever illustrative samples are needed so that the panel can do their job of describing the product well, letting the panel members discover the truth for themselves rather than parrot the leader's opinions.

A panel leader must have enthusiasm and a real liking for the job. Panelists must never feel that the sessions are boring. It follows that a panel leader must have a personality which assures that he or she can get along well with people. It helps a great deal in training a QDA® panel if people like the leader and trust his or her judgment.

Ideally, the panel leader should know the product being described. If the panel leader does not know the product well initially, he or she should have a good working knowledge of it by the time all the tasting and examining of the product necessary to set up the test series is done. It also helps if the panel leader has some knowledge or feel for what normal variability is with the standard or optimum product.

A good panel leader must have the ability to extract the essential idea from a great deal of detail since in both training and interpreting the data there are so many measurements and responses generated.

IX. RELATIVE STRENGTH VS. DIFFICULTIES OF USING THE QDA® TECHNIQUE

QDA® has many strengths which are attested to by those who use it on a continuing basis:

1. It gives very precise and reliable descriptive information which is specific for the product under examination.
2. Because the data can be statistically analyzed, the researcher can know whether the effects observed are significant or would have occurred by chance alone. Because the

data are replicated, it also provides measures of naturally occurring variability in the product or in the judge's performance.

3. QDA® information shows a high degree of repeatability over time if panelists are kept trained and motivated. This makes it possible to look at changes over time in a product.

4. The judges are relatively free from leader bias since they give their judgments in private and develop their own sensory terms and scoresheets.

5. QDA® data can be presented in a form which is communicable to the many different kinds of people who will have a use for it.

6. Finally, QDA® data can be used for many purposes. These have already been enumerated but include many types of problem solving and comparisons with competitive products; underlying measures of product quality, product quality monitoring, and interplant comparisons; determining the effect of packaging materials; and finally the development of new products and assessment of the effect of changes in old ones.

The difficulties of using the QDA® technique center around the time involved in screening and training the judges and, to a lesser extent, in conducting the replicated tests once the panel is trained. Because of the time involved, the technique is fairly costly. This appears to be a problem in some companies, but not in others.

Multiproduct companies also have problems keeping enough groups trained to cover a varied product line. This is sometimes solved by restricting the use of QDA® to important products or by training a single group of panelists to cover a related group of products.

Attrition of panel members from the trained groups either by in-house group member's retirement, promotion, or departure from the company, or by changing situations with outside judges is a problem, though this can be a problem with any descriptive technique. Turnaround time for data also can be a problem if analysis systems are not set up efficiently and staff-to-project ratios favor a lean headcount.

REFERENCES

1. **Stone, H., Sidel, J., Oliver, S., Woolsey, A., and Singleton, R. C.,** Sensory evaluation by Quantitative Descriptive Analysis, *Food Technol. (Chicago),* 28, 24, 1974.
2. **Anderson, N. J.,** Functional measurement and psychophysical judgement, *Psychol. Rev.,* 77(3), 153, 1970.
3. **Stevens, S. S.,** On the psychophysical law, *Psychol. Rev.,* 64, 153, 1957.
4. **Stone, H., Sidel, J. L., and Bloomquist, J.,** Quantitative descriptive analysis, *Cereal Foods World,* 25, 642, 1980.
5. **Stone, H. and Sidel, J. L.,** Descriptive analysis, in *Sensory Evaluation Practices,* Academic Press, Orlando, Fla., 1985, chap. 6.
6. **Caul, J. F.,** The profile method of flavor analysis, *Adv. Food Res.,* 7, 1, 1957.
7. **Brandt, D. A., Skinner, E. Z., and Coleman, J. A.,** Texture Profile Method, *J. Food Sci.,* 28, 404, 1963.
8. **Zook, K. and Wessman, C.,** The selection and use of judges for descriptive panels, *Food Technol., (Chicago),* 31, 56, 1977.
9. **Amerine, M. A., Pangborn, R. M., and Roessler, E. B.,** *Principles of Sensory Evaluation of Food,* Academic Press, New York, 1965, chap. 6.
10. **Cochran, W. G. and Cox, G. M.,** *Experimental Designs,* 2nd Ed., John Wiley & Sons, New York, 1957.
11. **Sidel, J. L., Stone, H., and Bloomquist, J.,** Use and misuse of sensory evaluation in research and quality control, *J. Dairy Sci.,* 64, 2296, 1981.

Chapter 4

BEER FLAVOR TERMINOLOGY — A CASE HISTORY

Morten C. Meilgaard

TABLE OF CONTENTS

I. Introduction 74
   A. Arguments for an Agreed Flavor Terminology 74
   B. History of the Development of the International
      Terminology for Beer 74

II. Basic Principles of the System 74

III. Description of the System 81

IV. Reference Substances 81

V. Use of the Terminology System and Reference Standards 86

References 86

I. INTRODUCTION

A. Arguments for an Agreed Flavor Terminology

The difficulty of communicating to another person the flavor notes present in a food or beverage is perhaps not appreciated. Whereas the basic standards of sight and hearing are taught in elementary school, those of smell and taste are acquired by chance. Semantic misconceptions abound; O'Mahony et al.[1] found that 1 out of 7 university students described citric acid as "bitter" rather than "sour", and 1 out of 12 students found quinine sulfate "sour" rather than "bitter". Once instructed, however, nearly all learned to correctly use the generally accepted terms.

The arguments for an agreed flavor terminology are the same as those for an agreed chemical terminology or biological terminology or those for a common scale of temperature. There are a number of important advantages in precise descriptions that have the same meaning to everyone suitably trained. For example, a brewer assessing the flavor effects of a process change must use the same terminology as the taste panel leader. Again, a chemist investigating a flavor defect must be able to make sense of the literature on the subject and must understand the flavor language not only of brewer and panel leader, but also of maltster, hop supplier, engineer, manager, foreman, etc. If any progress is to be made in the basic science of flavor chemistry, a chemist in one laboratory must make certain that flavor terms are defined in the same way as is done by colleagues in other laboratories.

B. History of the Development of the International Terminology for Beer

The efforts to establish the present system began in 1973 when a Joint Working Group was formed by the American Society of Brewing Chemists (ASBC), the European Brewery Convention (EBC), and the Master Brewers Association of the Americas (MBAA), with instructions to collect all terms in use and establish a set of basic principles which would permit selection of the best descriptors.

Early progress toward an international system is recorded in a 1975 report published in the brewing journals of several countries.[2] Discussions at the annual meetings of the collaborating organizations, extensive consultation, and voting by mail resulted (1977) in an intermediate system.[3] Proposals arising principally from experience in meeting practical brewing requirements eventually led to the third and, at the time of this writing, final form (Table 1), which was published in seven languages.[4] It was recognized at the outset that terminology is never static, nor should it be, and plans are in existence for a second working group to revise the system 10 or 20 years from this writing.

II. BASIC PRINCIPLES OF THE SYSTEM

The system is based on the following principles:

Each separately identifiable flavor characteristic has its own name — Although this means that there are well over 100 terms, the majority of the Joint Working Group felt that it would be wrong to simplify. Flavor is a complex phenomenon, and the system must contain enough terms to enable an expert taster to describe what is found. It must also provide suitable forms of terminology for all users, regardless of their level of training and experience.

Similar flavors are placed together — A user must be able to see at a glance the choice of terms. The order of terms should be the same in different languages. When a new term is introduced, it should be obvious where it belongs in the system.

No terms are duplicated for the same flavor characteristic — Ideally, both duplication and overlapping should be avoided, but in practice this leads to conflict with the first principle and with the requirement that all flavor characteristics must be covered. Language is an

Table 1
RECOMMENDED DESCRIPTORS

Class term	First tier	Second tier	Relevance[a]	Comments, synonyms, definitions	Reference standard
Class 1 — Aromatic, fragrant, fruity, floral					
	0110 Alcoholic		OTW	General effect of ethanol and higher alcohols	Ethanol, 50 g/ℓ
	0111 Spicy		OTW	Allspice, nutmeg, peppery, eugenol; see also 1003 Vanilla	Ethanol, 120 μg/ℓ
	0112 Vinous		OTW	Bouquet, fusely, wine-like	(White wine)
	0120 Solvent-like		OT	Like chemical solvents	
		0121 Plastics	OT	Plasticizers	
		0122 Can-liner	OT	Lacquer-like	
		0123 Acetone	OT		(Acetone)
	0130 Estery		OT	Like aliphatic esters	
		0131 Isoamyl acetate	OT	Banana, peardrop	(Isoamyl acetate)
		0132 Ethyl hexanoate	OT	Apple-like with note of aniseed; see also 0142 Apple	(Ethyl hexanoate)
		0133 Ethyl acetate	OT	Light fruity, solvent-like; see also 0120 Solvent-like	(Ethyl acetate)
	0140 Fruity		OT	Of specific fruits or mixtures of fruits	
		0141 Citrus	OT	Citral, grapefruit, lemony, orange rind	
		0142 Apple	OT		
		0143 Banana	OT		
		0144 Black currant	OT	Black currant fruit; for black currant leaves use 0810 Catty	
		0145 Melony	OT		(6-Nonenal, *cis-* or *trans-*)
		0146 Pear	OT		
		0147 Raspberry	OT		
		0148 Strawberry	OT		
	0150 Acetaldehyde		OT	Green apples, raw apple skin, bruised apples	(Acetaldehyde)
	0160 Floral		OT	Like flowers, fragrant	
		0161 2-Phenylethanol	OT	Rose-like	(2-Phenylethanol)
		0162 Geraniol	OT	Rose-like, different from 0161; taster should compare pure chemicals	(Geraniol)

Table 1 (continued)
RECOMMENDED DESCRIPTORS

Class term	First tier	Second tier	Relevance[a]	Comments, synonyms, definitions	Reference standard
		0163 Perfumy	OT	Scented	(Exaltolide musk)
	0170 Hoppy		OT	Fresh hop aroma; use with other terms to describe stale hop aroma; does not include hop bitterness (see 1200 Bitter)	
		0171 Kettle-hop	OT	Flavor imparted by aroma hops boiled in kettle	
		0172 Dry-hop	OT	Flavor imparted by dry hops added in tank or cask	
		0173 Hop oil	OT	Flavor imparted by addition of distilled hop oil	
Class 2 — Resinous, nutty, green, grassy					
	0210 Resinous		OT	Fresh sawdust, resin, cedar-wood, pine-wood, sprucy, terpenoid	
		0211 Woody	OT	Seasoned wood (uncut)	
	0220 Nutty		OT	As in brazil nut, hazelnut, sherry-like	
		0221 Walnut	OT	Fresh (not rancid) walnut	
		0222 Coconut	OT		
		0223 Beany	OT	Bean soup	(2,4,7-Decatrienal)
		0224 Almond	OT	Marzipan	(Benzaldehyde)
	0230 Grassy		OT		
		0231 Freshly cut grass	OT	Green, crushed green leaves, leafy, alfalfa	(*cis*-3-Hexenol)
		0232 Straw-like	OT	Hay-like	
Class 3 — Cereal					
	0310 Grainy		OT	Raw grain flavor	
		0311 Husky	OT	Husk-like, chaff, Glattwasser	
		0312 Corn grits	OT	Maize grits, adjuncty	
		0313 Mealy	OT	Like flour	
	0320 Malty		OT		
	0330 Worty		OT	Fresh wort aroma; use with other terms to describe infected wort (e.g., 0731 Parsnip/celery)	

Table 1 (continued)
RECOMMENDED DESCRIPTORS

Class term	First tier	Second tier	Relevance[a]	Comments, synonyms, definitions	Reference standard
Class 4 — Caramelized, roasted					
	0410 Caramel		OT	Burnt sugar, toffee-like	
	0411 Molasses		OT	Black treacle, treacly	
	0412 Licorice		OT		
	0420 Burnt		OTM	Scorched aroma, dry mouthfeel, sharp, acrid taste	
		0421 Bread crust	OTM	Charred toast	
		0422 Roast barley	OTM	Chocolate malt	
		0423 Smoky	OT		
Class 5 — Phenolic					
0500 Phenolic			OT		
		0501 Tarry	OT	Pitch, faulty pitching of containers	
		0502 Bakelite	OT		
		0503 Carbolic	OT	Phenol, C_6H_5OH	
		0504 Chlorophenol	OT	Trichlorophenol (TCP), hospital-like	
		0505 Iodoform	OT	Iodophors, hospital-like, pharmaceutical	
Class 6 — Soapy, fatty, diacetyl, oily, rancid					
	0610 Fatty acid		OT		
		0611 Caprylic	OT	Soapy, fatty, goaty, tallowy	(Octanoic acid)
		0612 Cheesy	OT	Dry, stale cheese, old hops	(Hydrolytic rancidity)
		0613 Isovaleric	OT		(Isovaleric acid)
		0614 Butyric	OT	Rancid butter (Hydrolytic rancidity)	Butyric acid. 3 mg/ℓ
	0620 Diacetyl		OT	Butterscotch, buttermilk	Diacetyl, 0.2— 0.4 mg/ℓ
	0630 Rancid		OT	Oxidative rancidity	
		0631 Rancid oil	OTM		
	0640 Oily		OTM		
		0641 Vegetable oil	OTM	As in refined vegetable oil	
		0642 Mineral oil	OTM	Gasoline (petrol), kerosene (paraffin), machine oil	
Class 7 — Sulfury					
0700 Sulfury			OT		
	0710 Sulfitic		OT	Sulfur dioxide, striking match, choking, sulfurous-SO_2	(KMS)

Table 1 (continued)
RECOMMENDED DESCRIPTORS

Class term	First tier	Second tier	Relevance[a]	Comments, synonyms, definitions	Reference standard
	0720 Sulfidic		OT	Rotten egg, sulfury-reduced, sulfurous-RSH	
		0721 H₂S	OT	Rotten egg	(H₂S)
		0722 Mercaptan	OT	Lower mercaptans, drains, stench	(Ethyl mercaptan)
		0723 Garlic	OT		
		0724 Lightstruck	OT	Skunky, sunstruck	
		0725 Autolysed	OT	Rotting yeast; see also 0740 Yeasty	
		0726 Burnt rubber	OT	Higher mercaptans	
		0727 Shrimp-like	OT	Water in which shrimp have been cooked	
	0730 Cooked vegetable		OT	Mainly dialkyl sulfides, sulfurous-RSR	
		0731 Parsnip/Celery	OT	Effect of wort infection	
		0732 DMS	OT	(Dimethyl sulfide)	DMS, 100 μg/ℓ
		0733 Cooked cabbage	OT	Overcooked green vegetables	
		0734 Cooked sweet corn	OT	Cooked maize, canned sweet corn	
		0735 Cooked tomato	OT	Tomato juice (processed), tomato ketchup	
		0736 Cooked onion	OT		
	0740 Yeasty		OT	Fresh yeast, flavor of heated thiamine; see also 0725 Autolysed	
		0741 Meaty	OT	Brothy, cooked meat, meat extract, peptone, yeast broth	
Class 8 — Oxidized, stale, musty					
0800 Stale			OTM	Old beer, over-aged, over-pasteurized	(Heat with air)
	0810 Catty		OT	Black currant leaves, ribes, tomato plants, oxidized beer	(p-Menthane-8-thiol-3-one)
	0820 Papery		OT	Initial stage of staling, bready (stale bread	(5-Methylfurfural, 25 mg/ℓ)

Table 1 (continued)
RECOMMENDED DESCRIPTORS

Class term	First tier	Second tier	Relevance[a]	Comments, synonyms, definitions	Reference standard
				crumb), cardboard, old beer, oxidized	
	0830 Leathery		OTM	Later stage of staling, often used in conjunction with 0211 Woody	
	0840 Moldy		OT	Cellar-like, leaf mold, woodsy	
		0841 Earthy	OT	Actinomycetes, damp soil, freshly dug soil, diatomaceous earth	(Geosmin)
Class 9 — Sour, Acidic					
		0842 Musty	OT	Fusty	
0900 Acidic			OT	Pungent aroma, sharpness of taste, mineral acid	
	0910 Acetic		OT	Vinegar	(Acetic acid)
	0920 Sour		OT	Lactic, sour milk; use with 0141 Citrus for citrus-sour	
Class 10 — Sweet					
1000 Sweet			OT		Sucrose, 7.5 g/ℓ
	1001 Honey		OT	Can occur as effect of beer staling (e.g., odor of stale beer in glass), oxidized (stale) honey	
	1002 Jam-like		OT	May be qualified by subclasses of 0140 Fruity	
	1003 Vanilla		OT	Custard powder, vanillin	(Vanillin)
	1004 Primings		OT		
	1005 Syrupy		OTM	Clear (golden) syrup	
	1006 Oversweet		OT	Sickly sweet, cloying	
Class 11 — Salty					
1100 Salty			T		Sodium chloride, 1.8 g/ℓ
Class 12 — Bitter					
1200 Bitter			TAf		(Isohumulone)
Class 13 — Mouthfeel					
	1310 Alkaline		TMAf	Flavor imparted by accidental admixture of alkaline detergent	(Sodium bicarbonate)

Table 1 (continued)
RECOMMENDED DESCRIPTORS

Class term	First tier	Second tier	Relevance[a]	Comments, synonyms, definitions	Reference standard
	1320 Mouthcoating		MAf	Creamy, onctueux (Fr.)	
	1330 Metallic		OTMAf	Iron, rusty water, coins, tinny, inky	(Ferious ammonium sulfate)
	1340 Astringent		MAf	Mouth puckering, puckery, tannin-like, tart	Quercitrin, 240 mg/ℓ[b]
		1341 Drying	MAf	Unsweet	
	1350 Powdery		OTM	O — Dusty cushion, irritating, (with 0310 Grainy) millroom smell Chalky, particulate, scratchy, silicate-like, siliceous	
	1360 Carbonation		M	CO_2 content	
		1361 Flat	M	Undercarbonated	60% of normal CO_2 content for the product
		1362 Gassy	M	Overcarbonated	140% of normal CO_2 content for the product
	1370 Warming		WMAf	See 0110 Alcoholic and 0111 Spicy	
Class 14 — Fullness					
	1410 Body		OTM	Fullness of flavor and mouthfeel	
		1411 Watery	TM	Thin, seemingly diluted	
		1412 Characterless	OTM	Bland, empty, flavorless	
		1413 Satiating	OTM	Extra full, filling	
		1414 Thick	TM	Viscous, epais (Fr.)	

[a] O = odor, T = taste, M = mouthfeel, W = warming, Af = afterflavor.
[b] Quercitrin is both astringent and bitter.

From Meilgaard, M. C., Dalgliesh, C. E., and Clapperton, J. F., *J. Am. Soc. Brew. Chem.*, 37, 48, 1979. With permission.

incomplete tool, providing a limited choice of words. For example, in five cases it was found necessary to permit overlapping pairs* of chemical name terms and generally descriptive terms. (To understand what this means, try to think of the totality of beer flavor as a three- or multidimensional continuum[5,6] in which related compounds are together and in which each compound and each term occupy a point or small volume in space. In such

* The five pairs are 0131 Isoamyl acetate and 0143 Banana, 0132 Ethyl hexanoate and 0142 Apple, 0133 Ethyl acetate and 0120 Solvent-like, 0613 Isovaleric and 0612 Cheesy, and 0732 DMS and 0734 Cooked sweet corn.

a system, 0613 Isovaleric and 0612 Cheesy would be close together but not coincident. 0613 Isovaleric would be closer to the area of the other fatty acids, and 0612 Cheesy would be more diffuse and would extend into the areas occupied by 0614 Butyric and by 0620 Diacetyl.)

The system is compatible with the *EBC Thesaurus* for the brewing industry.[7]

Subjective terms such as *good/bad, young/mature, or balanced/unbalanced* are not included — These hedonic terms have meaning within an active flavor panel and almost every brewery panel uses them, but they cannot be standardized on an international basis.

The meaning of each term is illustrated with readily available reference standards — See below.

III. DESCRIPTION OF THE SYSTEM

The system consists of 14 classes. These are given general names to indicate the types of flavors they contain. Only those terms preceded by a four-digit number are intended for use as descriptors. The remainder serve to indicate the class in which any given type of flavor should be sought. Some classes have a broader term (e.g., 0700 Sulfury) that serves as a common descriptor for all terms in the class; other classes do not have this because the language does not contain a suitable term.

There are three kinds of descriptors: class terms, first-tier terms, and second-tier terms. Broadly speaking, the first two are common terms familiar to most people, and together they form a vocabulary designed to fill most everyday needs.

The Flavor Wheel (Figure 1) is used to facilitate the location of these terms within the system. The wheel is meant as a memory aid and not as a separate system of classification of odors and tastes. Despite the diversity of terms in the system, a logical sequence could be maintained in most of it, but certain discontinuities do appear, as where 0700 Sulfury follows 0640 Oily. It can be argued that this occurs because a multidimensional continuum has been reduced to two dimensions for ease of presentation.

The second tier of terms in Table 1, together with the reference standards, serves the second purpose of the terminology, that of naming and identifying each separately identifiable flavor note in beer. These terms form the theoretical backbone of the system and are used mostly by specially trained panels. They also serve to define certain first-tier terms for which a reference compound or compounds are not available. For example, 0220 Nutty comprises a group of flavor notes exemplified by walnut-like, coconut-like, beany, and almond-like.

The column "Relevance" shows that most terms may be used to describe sensations of both odor (0) and taste (T). M, W, and Af indicate the terms that may be used to describe mouthfeel effects, warming, and afterflavor, respectively.

Under "Comments, synonyms, definitions" are given a number of terms that have been used in the past but should now be discouraged in favor of the more precise description given by the recommended terms. Thus, the term 0910 Acetic is preferred to "vinegar". The flavor note caused by caprylic and capric acids combined should be referred to as 0611 Caprylic.[8] The term 0630 Rancid is used only for oxidative rancidity (carbonyl compounds) and no longer includes any "butyric" flavor notes. The term "ribes" was discontinued and replaced by the more explicit and understandable 0810 Catty.

IV. REFERENCE SUBSTANCES

The Joint Working Group encountered a number of obstacles in its endeavor to choose readily available reference substances to illustrate each of the 122 terms of Terminology. Early "flavor libraries"[9-11] were presented in sniff bottles and consisted of solids, neat liquids, or substances dissolved in odorless nonyl phthalate or paraffin wax, but it soon

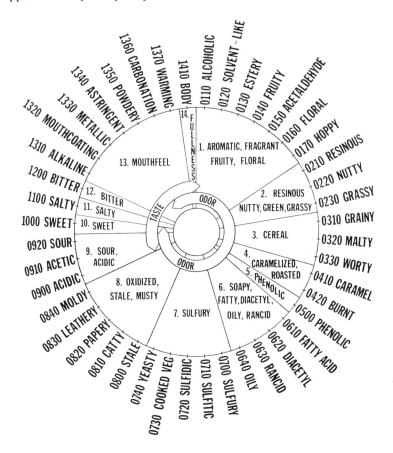

FIGURE 1.   Flavor Wheel showing class terms and first-tier terms. (From Meilgaard, M. C., Dalgliesh, C. E., and Clapperton, J. F., *J. Am. Soc. Brew. Chem.*, 37, 51, 1979. With permission.)

became apparent that in the training of panel members, sniff bottles, despite their convenience and reuseability, caused poorer recognition, more errors, and less precise flavor descriptions than did a system of standards dissolved in beer.

Once this was agreed upon, complaints appeared that several compounds were misflavored on receipt or had different flavors depending on the lot. A system of purification became neccessary and with this, a method of determining whether sensory purity has been attained.

Another obstacle not encountered to the same degree with sniff bottles was the need to find and use a narrow range of concentrations. For comparison, impressions of sight and hearing are instantaneous; the receptors can handle variations of six or seven orders of magnitude, and in testing situations any accidental overdose is easily toned down. The flavor impression, however, must reach and remain for some seconds at the gustatory or olfactory epithelium, which can handle at most two or three orders of magnitude, and sensory fatigue sets in rapidly; any overdose blinds the sensors for minutes or hours.

Visual or auditory aspects of an object can usually be separated in space or time, whereas a flavor note must generally be judged in a mixture with several other notes. As a result, instances have been recorded[12] in which panelists failed to detect a standard added at a concentration 10 to 100 times above their personal threshold because of unfamiliarity with its flavor. To be successful, therefore, a flavor demonstration requires a large amount of preparation, care, and subsequent control of what the subject perceived.

Any system of standards must account for the variations between individuals. Studies by the ASBC subcommittee[13] have shown that if panels are sufficiently large (e.g., 16 to 25 or more), the most sensitive 10% of panelists will have thresholds approximately four times below that of the median of the group, and the least sensitive 10% will have thresholds approximately five times above it. With the exception of anosmics, most healthy persons appear to show normal sensitivity for most substances, but each person tends to show high sensitivity for certain substances and low sensitivity for a few other substances.[14] No "supertaster" who had high sensitivity to all substances was found in any of the participating laboratories, nor was any pair of tasters discovered who showed exactly the same pattern of sensitivity.

An ideal reference compound was sodium chloride. It represents the term 1100 Salty quite perfectly. It is easily purified but can be used without purification. Stable and nonpoisonous, sodium chloride does not react with beer. It does not affect foam, color, or clarity, and it is soluble at 9 and even 81 times threshold concentration. Its flavor threshold is well defined, and it does not show a highly skewed or bimodal distribution of individual thresholds.

A second ideal reference compound was sucrose, for the term 1000 Sweet. Testing in these two cases could be limited to determinations of the threshold by the American Society for Testing and Materials (ASTM) Ascending Method of Limits.[12]

All other substances tested caused one or several problems that required separate investigation. With all little-known flavors, thresholds changed markedly with training so that three to five or more sittings of the panel were required before useful results could be obtained. Wide variations between tasters were encountered which necessitated repeated study of the compound by more than one panel, with each panel using the extended form of the ASTM method to detect and confirm tasters showing very high or very low thresholds, and to confirm or disprove any abnormal pattern of distribution of thresholds in the population which would make the compound in question a dubious choice as a reference standard.

The most difficult problems encountered were the removal of persistent impurities and the associated difficulty of determining whether a substance is sensorially pure. Criteria of chemical purity were of no help because impurities exist[15] which are flavor-active in the ppt range or even at the ng/ℓ level. The only viable alternative was "purification to constant flavor and threshold" in analogy with purification to a constant melting point.

An unexpected problem was that surface-active substances such as long-chain aliphatic compounds exist in variable states of solution.[16] When first added, they appeared to migrate to the surface of beer, causing a loss of foam, or they would cling to the wall of the container used. The problem was partly overcome by stipulating that (1) enough stock solution in ethanol/water be prepared so that all additions can be made directly into the bottles used for pouring taste tests and (2) bottles be stored for a minimum of 12 hr to allow time for complete solution of the added substances.

Reaction between beer and an added susbtance was seen in a few cases. Thresholds were found to be high as a consequence, as for H_2S and dimethyl sulfide in beers high in copper or iron.

The final recommendations of the Working Group are given in Table 2 which also contains the threshold found for each standard, expressed as a range in beers containing various levels of the compound or flavor in question. For example, the threshold for dimethyl sulfide is given as "25 to 50 $\mu g/\ell$ in beer containing 10 to 100 $\mu g/\ell$". By interpolation, if a given beer contains 50 $\mu g/\ell$, then the group threshold can be expected to fall to near 35 $\mu g/\ell$.

The use of reference substances is described below, and full details of the methods of purification, together with the outcomes of practical brewery tests, are discussed by Meilgaard et al.[17]

Table 2
FLAVOR REFERENCE SUBSTANCES

Term		Compound	Supplier	Method of purification	Difference threshold[a]	Content in beer
0110	Alcoholic	Ethanol	High-quality vodka[b]	None required	17 g/ℓ	33—42 g/ℓ
0111	Spicy	Eugenol	Aldrich	Solvent wash + fractional distillation + adsorption	40 µg/ℓ	—
0131	Isoamyl acetate	Isoamyl acetate	Aldrich	Adsorption + GC	0.5—1.7 mg/ℓ	1—3 mg/ℓ
0132	Ethyl hexanoate	Ethyl hexanoate	K & K Laboratories	Adsorption + GC	0.15—0.25 mg/ℓ	0.2—0.4 mg/ℓ
0133	Ethyl acetate	Ethyl acetate	Fluka	Adsorption	20—40 mg/ℓ	10—30 mg/ℓ
0145	Melony	Melonal®[c]	Givaudan	None required	1 µg/ℓ	—
0150	Acetaldehyde	Acetaldehyde	Merck	Adsorption + distillation + adsorption	10—20 mg/ℓ	2—10 mg/ℓ
0162	Geraniol	Geraniol	Merck	Use fresh supply	~150 µg/ℓ[d]	0—60 µg/ℓ
0173	Hop oil	Cluster hop oil[e]	S. S. Steiner	None required	0.1 mg/ℓ	—
0224	Almond	Benzaldehyde	Aldrich	None required	1 mg/ℓ	—
0611	Caprylic	Octanoic acid	Sigma	Recrystallization of calcium salt	5—10 mg/ℓ	2—8 mg/ℓ
0613	Isovaleric	Isovaleric acid	Sigma	None required	0.5—1.5 mg/ℓ	0.5—1.5 mg/ℓ
0614	Butyric	Butyric acid	Merck	2 × Fractional distillation	2—3 mg/ℓ	0.5—1.5 mg/ℓ
0620	Diacetyl	2,3-Butanedione	Aldrich	Fractional distillation + adsorption	0.07—0.15 mg/ℓ	0.03—0.3 mg/ℓ
0710	Sulfitic	Sodium metabisulfite	Fisher Scientific	None required	20 mg/ℓ SO_2	1—10 mg/ℓ SO_2
0721	H_2S	Sodium sulfide	Mallinckrodt	Select colorless crystals	4 µg/ℓ H_2S	0—2 µg/ℓ
0722	Mercaptan	Ethanethiol	Aldrich	None required	1 µg/ℓ	0—0.5 µg/ℓ
0732	DMS	Dimethyl sulfide	Matheson, Coleman and Bell	Adsorption	25—50 µg/ℓ	30—100 µg/ℓ
0841	Earthy	Geosmin	National Environment Research Center	None required	0.1 µg/ℓ	—
0841	Earthy	2-Ethyl fenchol	PFW, Inc.	None required	5 µg/ℓ	—
0910	Acetic	Acetic acid	J. T. Baker "Ultrex"	None required	60—120 mg/ℓ	30—200 mg/ℓ
1000	Sweet	Sucrose	Grocery	None required	2.6 g/ℓ	—

1003	Vanilla	Vanillin	Fluka	None required	40 $\mu g/\ell$	0—10 $\mu g/\ell$
1100	Salty	NaCl	Grocery	None required	0.6 g/ℓ	—
1200	Bitter	Isohumulone	Kalsec "Isolone"ᶜ	None required	7—15 mg/ℓ	0—30 mg/ℓ
1330	Metallic	$FeSO_4 7H_2O$	J. T. Baker	None required	1 mg/ℓ Fe	0—0.5 mg/ℓ
1340	Astringent	Quercitrinᵍ	K & K Laboratories	Recrystallization from 50% ethanol	80 mg/ℓ	—

[a] The standard recommended addition for reference purposes is three times the threshold.

[b] Smirnoff or equivalent. Strength varies with locality and the vodka must be analyzed before use. Addition to beer should be by weight, not by volume.

[c] Trade name for 2,6-dimethyl-5-hepten-1-al. Store under refrigeration.

[d] Thresholds of geraniol added to beer show a distribution with two maxima at 18 $\mu g/\ell$ (35% of persons studied) and 350 $\mu g/\ell$ (65%). Recommended addition for reference purposes = 1 mg/ℓ.

[e] Not a reference standard; recommended for demonstration purposes.

[f] A solution of varying strength, usually 17%.

[g] Quercitrin is both astringent and bitter.

From Meilgaard, M. C., Reid, D. S., and Wyborski, K. A., *J. Am. Soc. Brew. Chem.*, 40, 121, 1982. With permission.

V. USE OF THE TERMINOLOGY SYSTEM AND REFERENCE STANDARDS

In practical use, a panel manager putting together a scoresheet for descriptive analysis chooses from Table 1 those terms that are appropriate for any given test; others (such as hedonic terms) may be added if required. For example, in daily tasting of one brewery's beer, scoresheets commonly have 10 to 20 terms, whereas two or three terms may be too many for certain consumer tests. On the other hand, if for example a sulfury off-flavor develops, a form that has all of the 18 terms under 0700 Sulfury may be appropriate. The terminology is designed to be sufficiently flexible, yet self-consistent, to meet all these requirements.

Reference substances are used in the training of tasters as follows:

1.   A stock solution of the standard is prepared in water, ethanol, or aqueous ethanol of a concentration such that neither less than 5 $\mu\ell$ (the minimum for accuracy using a microsyringe) nor more than 500 $\mu\ell$ of ethanol (the maximum before ethanol flavor becomes apparent) needs to be added per 355 mℓ-bottle of null beer. A relatively bland beer that is well known to the panelist should be used as null beer.

2.   Bottles should be prepared for tasting by injecting the required concentration of stock solution below the surface of the beer, which should have been cooled to 4 to 10°C before uncrowning. The bottle should be tapped by a metal object until a head of foam expels the air in the neck (but not the added substance). The bottle should be recrowned and stored at 4°C overnight before tasting.

3.   The panelist should be given a glass of beer to which is added the demonstration standard in a concentration nine times the threshold (Table 2). The glass (or the bottle from which it is poured) should be clearly marked with the name of the standard followed by "9 times threshold". The panelist is to state the strength and type of added flavor perceived and to discuss the flavor until it is clear that it is understood. In rare cases, the panelist may not perceive the added flavor because of low sensitivity, in which case concentrations of "81 times threshold" may be tried.

4.   The panelist should be given a triangle test comparing the demonstration standard blindly in a concentration of three times threshold against the untreated control beer. The panelist should be regarded as having understood the flavor if the scoring is correct on both the second and third such triangles. A miss on one of these should cause either a return to step 3 or, if the panelist shows lack of interest, disqualification with regard to this particular flavor.

REFERENCES

1. **O'Mahony, M., Goldenberg, M., Stedmon, J., and Alford, J.,** Confusion in the use of the taste adjectives "sour" and "bitter", *Chem. Senses Flavor,* 4, 301, 1979.
2. **Clapperton, J. F., Dalgliesh, C. E., and Meilgaard, M. C.,** Progress towards an international system of beer flavor terminology, *Tech. Q. Master Brew. Assoc. Am.,* 12, 273, 1975.
3. **Clapperton, J. F., Dalgliesh, C. E., and Meilgaard, M. C.,** Systematic beer flavor terminology, in *The Practical Brewer,* Broderick, H., Ed., Master Brewers Association of the Americas, Madison, Wis., 1977, 433.
4. **Meilgaard, M. C., Dalgliesh, C. E., and Clapperton, J. F.,** Beer flavor terminology, *J. Am. Soc. Brew. Chem.,* 37, 47, 1979.
5. **Clapperton, J. F.,** Sensory characterization of the flavour of food and extracts, in *Progress in Flavour Research,* Land, D. G. and Nursten, H. E., Eds., Applied Science, London, 1979, 1.
6. **Moskowitz, H. R. and Gerbers, C. L.,** Dimensional salience of odors, *Ann. N.Y. Acad. Sci.,* 237, 1, 1974.

7. EBC Information and Documentation Group, EBC Thesaurus, 2nd ed., Vols. 1 and 2, European Brewery Convention, Zoeterwoude, Neth., 1983.
8. **Clapperton, J. F. and Brown, D. G. W.,** Caprylic flavour as a feature of beer flavour, *J. Inst. Brew.,* 84, 90, 1978.
9. **Harper, R.,** Some chemicals representing particular odour qualities, *Chem. Senses Flavor,* 1, 353, 1975.
10. **Clapperton, J. F.,** *European Brewery Convention, Proceedings of the 15th Congress,* Elsevier, Amsterdam, 1975, 823.
11. **Williams, A. A.,** The development of a vocabulary and profile assessment method for evaluating the flavor contribution of cider and perry aroma constituents, *J. Sci. Food Agric.,* 26, 567, 1975.
12. American Society of Brewing Chemists, Report of subcommittee on sensory analysis, *J. Am. Soc. Brew. Chem.,* 38, 99, 1980.
13. **Brown, D. G. W., Clapperton, J. F., Meilgaard, M. C., and Moll, M.,** Flavor thresholds of added substances, *J. Am. Soc. Brew. Chem.,* 36, 73, 1978.
14. **Meilgaard, M. C. and Reid, D. S.,** Determination of personal and group thresholds, and the use of magnitude estimation in beer flavour chemistry, in *Progress in Flavour Research,* Land, D. G. and Nursten, H. E., Eds., Applied Science, London, 1979, 67.
15. **Meilgaard, M. C.,** Flavor chemistry of beer. II. Flavor and threshold of 239 aroma volatiles, *Tech. Q. Master Brew. Assoc. Am.,* 12, 151, 1975.
16. **Roberts, R. T. and Clapperton, J. F.,** Flavour of fatty acids in relation to their physical state in solution, *J. Inst. Brew.,* 84, 157, 1978.
17. **Meilgaard, M. C., Reid, D. S., and Wyborski, K. A.,** Reference standards for beer flavor terminology system, *J. Am. Soc. Brew. Chem.,* 40, 119, 1982.

Chapter 5

THE TEXTURE PROFILE METHOD

Elaine Z. Skinner

TABLE OF CONTENTS

I. Introduction 90

II. Review of the Texture Profile Development and Applications 90

III. Rating Scales Used in Texture Profiling 95

IV. Panel Training 96

V. Texture Vocabulary 98

VI. Data Analysis 98

VII. Application of the Texture Profile to Fat-Containing Foods 99
    A. Fats and Oils 99
    B. Fat-Containing Foods 100
    C. Creaminess of Foods 101
    D. Application to Cheesecakes 103

VIII. Overview 103

References 107

I. INTRODUCTION

The General Foods (G.F.) Texture Profile Method, its origin, and its development were described by Brandt et al.[1] in 1963 and revisited by Szczesniak in a 10-year perspective in which she describes its use by researchers here in the U.S. as well as abroad.[2] This chapter updates developments and applications of the Texture Profile Method and reviews the method since its inception.

Researchers from many fields have contributed their thoughts and ideas about this method in various publications and communications throughout the past 20 years. Bourne et al.[3] applied the Texture Profile in another cultural environment through the training of panels and panel leaders in Colombia. He also constructed new standard scales composed of foods available in the Colombian marketplace, confirming that these scales are amenable to modification as Szczesniak and Civille and Szczesniak pointed out.[2,4]

Relationships among texture profile panel results, consumer results, and objective measurements were discussed by Cardello et al.[5] and Moskowitz et al.[6] Moskowitz and Kapsalis pointed to the historical fit of the Texture Profile with sensory and perceptual psychology and the developing science of sensory evaluation.[7]

Arocha and Toledo, Pangborn and Koyasako, and Ingate and Christensen showed applications of the Texture Profile or extensions of the approach to foods and beverages.[8-10] The adaptation of the Texture Profile Method to skin care products was reviewed by Schwartz,[11] while Civille and Liska reported on modifications made in the Texture Profile which expanded upon the original work.[12] Civille and Szczesniak developed a specific set of guidelines for training a texture panel to use the G.F. Texture Profile system.[4]

This chapter provides an historical perspective by:

1. Reviewing applications of the Texture Profile Method to a wide variety of foods
2. Reviewing the expansion/revisions made in the methodology and procedures
3. Discussing some of the components of the method (rating scales, training concepts, and vocabulary) in greater detail than shown in the original papers

It also covers more recent developments, i.e., expansion and application to fat-containing foods.

II. REVIEW OF THE TEXTURE PROFILE DEVELOPMENT AND APPLICATIONS

Table 1 shows in historical perspective the development and applications of the Texture Profile at General Foods. In the 20 years since its development, the method has been applied to a broad range of food products covering solid, semisolid, and liquid systems. It has also been adapted to the evaluation of cosmetics.

As can be seen in Table 1 the methodology was developed in the early to mid-1960s and was applied to cereals, quick-cooking rice, cookies, candy, and strawberries. All of these products represent solid systems which were easily handled within the limited scope of the methodology at that time.

In the latter part of the 1960s, research in emulsion technology led to the development of whipped topping systems which culminated in the development of Cool Whip® and Dream Whip®. Sensory texture evaluation played a key role in these product development activities. Building on the original work, terminology and evaluation techniques were developed to address semisolid systems.[12] The emerging relationships of sensory descriptions of texture with instrumental measures encouraged exploration into developing consumer testing methodology for assessing consumer awareness of texture. The Consumer Texture Profile Technique grew out of this investigation.[13]

Table 1
DEVELOPMENTS IN SENSORY TEXTURE PROFILING

1960—1965
Development of Methodology: Original Work

Texture characteristics grouped in three main classes:
Mechanical
  Primary parameters
    Cohesiveness
    Elasticity
    Hardness
    Viscosity
    Adhesiveness
  Secondary parameters
    Brittleness
    Chewiness
    Gumminess
Geometrical
  Characteristics related to size and shape
  Characteristics related to shape and orientation
Other
  Moisture content
  Fat content
    Oiliness
    Greasiness
Applications (primarily solids)
  Ready-to-eat cereals
  Instant rices
  Cookies
  Candy
  Strawberries

1965—1970
Expansion of Original Work

Semisolids
  Techniques established for standardizing manipulation of semisolids in the mouth
  Terminology established for describing semisolid texture characteristics, i.e., reaction to manipulation of tongue,
    reaction to saliva, and rate and type of product disappearance
Chewing gums
  Techniques for standardizing the mechanics of chewing established
  Nomenclature developed for describing mastication over 10 min
  Comprehensive nomenclature developed for describing gum texture characteristics
Development of consumer Texture Profile technique
  Texture terms adapted for consumers
  6-Point intensity scale selected
  Consumer questionnaire developed
Applications (semisolid/gums)
  Whipped toppings
  Puddings and flans
  Butter/margerine
  Gelatin desserts
  Stick gums
  Bubble gums
Other applications
  Potato chips
  Pancake mixes
  Moisture mimetic foods
  Intermediate moisture food

Table 1 (continued)
DEVELOPMENTS IN SENSORY TEXTURE PROFILING

1970—1975
Nonfood Application, Changes/Additions to Original Method

Skin care products — methodology developed
  Evolved as combination of specific mechanical characteristics. Evaluation in three stages:
    1. Pick-up — removal of product from container
    2. Rub-out — application of product to skin
    3. After-feel — evaluation of effect of product on skin
Changes/additions to original work
  Evaluation begins with surface properties rather than first bite or compression
  Evaluation of ''cohesiveness of the mass'' is initiated to account for characteristics of the mass when mixed
    with saliva
  Elasticity changed to springiness due to confusion of term with rheological and physical definitions
  Brittleness changed to fracturability which is considered to reflect entire scale more clearly
Applications
  Skin care and cosmetic products
  Macaroni
  Sausage
  Hot dogs
  Stuffing mixes
  Cob corn
  Chip dips
  Chicken
  Ham

1975—1980
Further Expansion of Original Work

Liquids
  Technique and terminology established for evaluation of beverages and beverage mouthfeel
Fats and oils
  Technique and terminology developed for evaluation of fat-containing and low fat products
Applications (liquids/fats and oils/carbonation)
  Peanut butter
  Sour cream
  Cheese cakes
  Mayonnaise
  Cream cheese
  Fats and oils
  Powdered soft drinks
  Carbonated beverages
  Pop rocks or carbonated candy
Other applications
  French fries
  Fish fillets
  Shellfish
  Bacon
  Roast beef

1980—1985
Developed Techniques for Variety of Products

Frozen desserts
  Emphasis given to mouthfeel properties, temperature effects, melting properties
Syrups
  Special attention given to techniques for sensory evaluation of viscosity
    applications (desserts/syrups)
  Syrups
  Soft ice cream
  Hard ice cream
  Frozen confections

Table 1 (continued)
DEVELOPMENTS IN SENSORY TEXTURE PROFILING

Other applications
  Coating mixes
  Brie
  Cheeses
  Breads
  Frosting

Table 2
TEXTURE PROFILE OF TWO RICE SAMPLES[1]

	Rice R	Rice Q
Initial		
Mechanical characteristics		
Hardness (1—9 scale)	2.3	2.7
Geometrical characteristics	None	None
Masticatory		
Mechanical characteristics		
Chewiness (1—7 scale)	1.3 (12 sec.)	1.7 (15.0 sec.)
Geometrical characteristics	Two-phase system — outside phase slimy and soft, inside phase gelatinous and rubbery	Slightly mealy
Residual		
Rate and type of breakdown	Type of breakdown is uneven due to heterogeneous structure. Rate of breakdown is fairly rapid because of relatively low hardness and chewiness	Even and slow breakdown. Kernel texture is uniform, slightly mealy and dry after breakdown

The opportunity to broaden experience with solid food evaluation was provided through the assessment of potato chips, pancake mixes, and moisture-managed food.

During the 1970s, there was even broader expansion and application of the Texture Profile. Sensory methodology for the evaluation of skin care products was developed by adaptation of the Texture Profile principles and methodology.[11]

The original work did not cover the surface properties of foods (e.g., smoothness, wetness, chalkiness, greasiness, etc.) perceived by the soft tissues in the mouth prior to biting and chewing. Because of their potential importance to product quality and consumer acceptance, methodology and procedures were developed for evaluating these properties.[4,12]

A good example of the evolution of the Texture Profile method can be seen in the texture profiles of rice reported in the original paper and in Civille's paper on panel training guidelines.[1,4] These are reprinted in Tables 2 and 3, respectively, for ease of comparison.

Contrast the paucity of information about the textural differences between the two rice samples in Table 2 with those shown in Table 3. Table 3 describes differences in the surface properties of wetness, kernel stickiness, roughness, uniformity of size, clumpiness, and plumpness among the rice samples (Stage 1). This information is lacking in Table 2. Again, in Table 3, not only is the mechanical characteristic of hardness evaluated quantitatively, but so are the parameters of crumbliness, rubberiness, gluiness, and inner moisture.

In Table 2, Stage 3 or residual characteristics is a description of rate and type of breakdown only, whereas kernel uniformity, cohesiveness of the mass, and "sticks to teeth" are quan-

Table 3
DETAILED SENSORY TEXTURE PROFILE OF SEVERAL TYPES OF RICE

	Brand			
	A	**B**	**C**	**D**
Stage 1				
Wetness	2	2—3	2	1—2
Kernel stickiness	1—2	)(-1	)(	)(
Roughness	1—2	)(-1	1—2	2-3
Uniformity of size	1—2	2—3	3	1—2
Clumpiness	2—3	)(-1	1	0-)(
Plumpness	1—2	2—3	3	)(-1
Stage 2				
Hardness	1—2	1—2	1—2	1—2
Crumbliness	)(-1	)(	0	1—2
Rubberiness	)(-1	1—2	2—3	)(-1
Gluiness	2—3	1—2	)(-1	)(
Inner moisture	1	2	1—2	1
Stage 3				
Kernel uniformity	1	1—2	)(-1	2
Cohesiveness of mass	2—3	1—2	)(	1
Sticks to teeth	1—2	1	0	1—2
Mouthcoating	Thin, starchy	Chalky	Slightly chalky	Throat coating
Breakdown rate	Fast—moderate	Fast—moderate	Slow—moderate	Slow—moderate
Breakdown type	Breaks into soft, grainy particles in a pasty slurry	Breaks into distinct beadlike particles	Breaks into soft grainy pieces	Crumbles into gritty pieces

Note: An explanation of the ratings is found in Table 4.

From Civille, G. V. and Szczesniak, A. S., *J. Texture Stud.*, 4(2), 204, 1973. With permission.

tified in Table 3. The concept of "cohesiveness of the mass" was added to facilitate the evaluation of mass formed during the chewing process as the result of saliva incorporation with the product.[12]

Some minor terminological changes were also made at this time. Elasticity was renamed springiness to lessen the apparent confusion caused by conflict with rheological definitions. Brittleness was renamed fracturability because the latter was felt to reflect more clearly the reference products of the standard scale.[2] Other products covering a broad range of texture types such as cob corn, chicken, macaroons, and chip dip, among others, were evaluated at this time.

During the late 1970s, evaluation techniques and vocabulary were established for evaluation of beverage mouthfeel and fat-containing foods. The application to fat-containing foods is described in detail in a latter section of this chapter. Other product applications were to deep-fat-fried foods (potatoes and fish), shellfish, and meat analogs.

During the 1980s, techniques were developed for the evaluation of frozen dessert products and syrups. In the former case, emphasis was placed on mouthfeel properties, temperature effects, and melting properties. In the latter case, special attention was given to techniques for evaluation of syrup viscosity. Other product applications were to coating mixes, cheeses, breads, and frostings.

The Texture Profile continues to be widely used within General Foods, both domestically and abroad, to evaluate a variety of products for a variety of research and development (R & D) objectives.

III. RATING SCALES USED IN TEXTURE PROFILING

Rating scales in general have created ongoing, interesting arguments among practitioners and theoreticians alike and will probably continue to do so for many years to come. The rating scales of the Texture Profile have received their fair share of comment from the technical community, including some of the originators of the system.

Many who have attempted to use the Texture Profile in their laboratories have asked for clarification and explanation of the concept and use of the rating scales. Szczesniak briefly addressed this issue, explaining that two scales exist — the standard rating scales for quantifying the mechanical parameters of texture and an intensity scale for evaluating test samples.[2] Civille and Szczesniak elaborated on when and how to use the two scales.[4]

Because of the importance of scaling to the successful use of the method, it bears further discussion. The Standard Rating scales of hardness, brittleness, chewiness, gumminess, viscosity, and adhesiveness were designed to cover the entire intensity range of each of these parameters found in food products. Also, they were intended to serve the panel with a well-defined, comprehensive frame of reference for learning about and assessing food textures and were selected so adjacent scale points were equidistant from each other.

The scale of hardness, for example, consists of nine food product references that illustrate varying degrees of hardness ranging from cream cheese, which has a low degree of hardness, to rock candy, which has a high degree of hardness. The lowest scale point was assigned a value of 1 and the highest a value of 9.

In selecting the lowest degree of hardness, it was found that cream cheese was the best available commercial reference that met the criteria of being a predominant characteristic of the food, being present at an appropriate intensity (in this case low hardness), and easy to perceive sensorially. On the upper end of the scale, few commercial food products that were harder than rock candy also fulfilled the other previously mentioned criteria.

In a like manner, scales were developed for the other mechanical parameters, the exception being the gumminess scale in which flour/water pastes were used instead of commercial foods. The attempt was made at that time to use the scale values of the standard scales as intensity values for quantifying samples.

The original paper on the Texture Profile shows profiles of rice samples, whipped toppings, and biscuits.[1] The reader will notice that the mechanical parameters (hardness, chewiness, and bitterness) applicable to these products were quantified using the standard rating scale values.

However, it was found very quickly that unless samples differed from each other substantially, such as stale and fresh biscuits, the distances between points on the standard scales were often too large to reflect the smaller but discernible sample differences typically encountered in product R & D. Also, it was more difficult for panel members psychologically to use the values of the standard rating scales as scores for quantifying differences among test samples.

Consequently, a 14-point intensity scale was adopted which is an expanded version of the)(-3 Flavor Profile scale. The scale is printed in its entirety in Table 4. This scale was selected since the panel members who comprised the first Texture Profile panel were experienced with this 14-point scale from flavor testing.

During the training of a texture profile panel, a multistage approach is taken to teaching the use of the rating scales. First, the panel learns to rank-order texture characteristics through repeated exposure to each standard rating scale. Next, panelists learn to scale differences on a three-point scale representing small, moderate, and large differences. With additional practice and exposure to a larger population of test samples, they eventually use the full range of the 14-point intensity scale.

Other category scales, magnitude estimates, and line scales can also be used as the means

Table 4
TEXTURE PROFILE
INTENSITY SCALE

Rating[a]	Definition
0	None
)(	Barely perceptible
)(-1	
)(-1	
)(-1	
1	Slight
1—2	
1—2	
1—2	
2	Moderate
2—3	
2—3	
2—3	
3	Strong

[a] Underline indicates a value between the two numbers indicated, but closer to the underlined number.

of data quantification.[14,15] The main criterion is that the panel be familiar with using the scale in a variety of situations. A recent study compared the 14-point intensity scale shown in Table 4, a 6-in. line scale, and a fixed-modulus magnitude estimation scale. Panelists who participated in this study had equal experience with the three scaling methods. The product vehicle was pound cakes. Results showed that all three scales performed equally well in differentiating among the products.[16]

In summary, practitioners are advised to use the standard rating scales for teaching and demonstration purposes; for product testing, an intensity rating scale of choice should be used to quantify texture characteristics of samples.

IV. PANEL TRAINING

Training of panel members is an essential component of descriptive analysis techniques. Cairncross and Sjostrom,[17] Brandt et al.,[1] and Stone and Sidel[18] discuss panel training for the Flavor Profile, Texture Profile, and the QDA® Method respectively.[18] Civille and Szczesniak[4] describe in detail the procedures to be followed for training a Texture Profile panel.

The question of what constitutes training needs to be further addressed. There are probably as many definitions of a trained panel as there are sensory practitioners using trained panels. The sensory literature is vague for the most part. Typically, authors refer to having used a trained panel to evaluate the products being discussed without providing any additional information about the training of the panel.

Perhaps communication on the subject of training could be improved if the words "panel member development" were substituted for "panel training" and further identified and defined factors that constitute the stages of panel member development:

- Recruitment — enrolling candidates for screening

Table 5
PERCENTAGE OF TIME
SPENT IN TEXTURE
PROFILE PANEL
MEMBER
DEVELOPMENT

Task	Time (%)
Recruitment	15
Screening/selection	8
Training	23
Practice	46
Analysis/report	8

● Screening — assessing candidate performance on the basis of appropriate physiological tests, availability, and interest

● Selection — choosing candidates for panel training and eventual panel membership based on meetng minimum criteria established by screening

● Training — educating candidates in the qualitative and quantitative aspects of a sensory method and sample evaluating techniques through a series of lecture/demonstration sessions.

These four stages of panel member development are part of the Texture Profile system because this ''front-end loading'' is believed to make the difference between an effective panel and one that needs frequent returns for tune-ups.

Table 5 shows a break-out of time spent on each of these stages. As can be seen in the table, the major portion of time (69%) is invested in training (23%) and practice (46%).

The ''guts'' of a good descriptive panel system are in the training and practice portion of the program. This includes the amount, as well as the type and quality, of the training and practice material and the substantive knowledge and experience of the instructors. It is not sufficient to be a good process leader in the training of a descriptive panel. An analogy can be drawn between teaching students to learn a language and teaching a group of trainees to learn texture profiling. If the instructor is not proficient in using the language, the students will not become proficient. At best, they might be able to communicate on a superficial, trivial level; at worst, they might not be able to communicate at all.

During the practice program which follows immediately after the training course, the panelists begin to develop their muscles, so to speak, and learn to become confident and competent in evaluating products. It is during this time that they begin to develop their memory banks of vocabulary/product associations. It is also during this time that a data base is being developed on each panelist with respect to individual and group data repeatability.

Relatively little time (8%) is spent on candidate screening/selection. This in no way minimizes the importance of this stage of panel member development. However, the screening of candidates requires only sufficient time to expose them to the sensory testing environment, particularly if they have had no prior sensory test experience. This is done by introducing them to relevant test materials and tasks and through an interview, determining their availability status and their interest/attitudes about becoming a panel member. When physiological testing is a requirement, e.g., assessment for taste blindness and anosmia for flavor work, it is made part of the screening process.

For texture, no physiological criteria currently exist for rejecting candidates. Civille describes a simple test that can be used during screening for texture candidates. Four reference products from the standard scale of hardness are presented in order and candidates are asked to arrange them in increasing order of hardness.[4] This test satisfies the requirement of exposing candidates to the sensory testing environment while also providing some data feedback. Artificial denture wearers, however, are excluded from texture training.

Typically, texture panel candidates are also screened for flavor training because panelists are expected to participate in both flavor and texture testing, and it is efficient to screen for both types of panels at the same time. Although texture training is conducted separately from flavor training, texture and flavor profiles are developed on products in actual product test situations.

Although not correct in their understanding of the Texture Profile application, Stone and Sidel are quite right in saying that "separation of texture from other sensory properties of a product such as color, aroma, taste, etc. is a concern".[18]

There are many different opinions, interpretations, and practices regarding descriptive panel selection criteria. The two main camps tend to be those who champion intensive and extensive multiple-product training and those who champion intensive product-specific screening. The Flavor and Texture Profile methods are illustrative of the former and the QDA® Method is illustrative of the latter. Practitioners tend to rely on their favorite selection scenarios and consequently have little objective comparative data to support their choices.

V. TEXTURE VOCABULARY

Vocabulary is the cornerstone of the Texture Profile. Consequently, considerable time and effort are devoted to it both in panel training and in applications to products.

Detailed substantive coverage of vocabulary can be found in the papers by Brandt et al.,[1] and Szczesniak et al.[19] The process aspects of vocabulary development have been given comparatively little coverage and thus are addressed here together with a brief review of the sensory definitions of texture.

Terminology which correctly defined the rheology of food texture had to be revised in order for it to make sensory sense. For example, hardness, which was defined rheologically as the force required to deform, was redefined as the force required to penetrate a substance with the molar teeth. Likewise, sensory definitions were developed for all of the other mechanical parameters of texture. As correctly pointed out by Stone and Sidel, "The technical nature of these (rheological) definitions should be meaningful to the chemist, but one might question their perceptual meaning to a test subject."[18]

The texture references available in the form of the standard rating scales and examples of geometrical, fat, and moisture characteristics provide the panel with an experiential knowledge base. This knowledge base forms a pathway through an otherwise uncharted jungle of possible terms, facilitates communications, and expedites the development of texture profiles on product submissions in actual test situations.

Those products that are unfamiliar to the panel in a testing context are first examined in orientation sessions. These sessions usually consist of an exposure to the "world" of the product category, viz., several marketed brands and available experimental prototypes.

The panel initially evaluates this product without consulting reference materials and without a formal ballot. At this time, the objective is to sort through the products, categorizing them in terms of similarities and differences.

When this process is completed, the panel then begins development of an appropriate texture vocabulary using the standard texture scales as well as all other reference materials that will aid in describing the product of interest. In short, the panels do not work for the method but rather let the method work for them.

VI. DATA ANALYSIS

There is no standard procedure for analyzing texture profile data. In practice, analysis methods vary from panel consensus to multivariate statistical analyses. There is increasing use of multivariate analyses applications to descriptive sensory data, principally because the increasing use of computers facilitates multivariate analyses. The state of the art is covered well in Powers' paper on the evaluation of sensory data by statistical programs.[20] The advantages offered by these programs are many, from tracking panel member performance to data reduction and data display.

However, these statistical procedures are also being misused by their application to test method simplification. Syarief et al.[21] have used Principal Components Analysis "to study the interdependency and the underlying dimensions of the sensory textural characteristics of products in an attempt to simplify the method of texture profile analysis". The authors concluded that "the texture profile analysis may be simplified by reducing character notes being evaluated to about half or less of the number of original notes".[21] This implies a fundamental lack of understanding about sensory testing in general and about descriptive techniques in particular. The multivariate methods are best used as data reduction tools. To use these or any other tools to "simplify" Texture Profile and other forms of sensory descriptive analysis is an abuse of the methods, as well as an inappropriate use of the statistical techniques.

VII. APPLICATION OF THE TEXTURE PROFILE TO FAT-CONTAINING FOODS

The impetus for exploring the sensory contribution of fat content to foods grew out of interest in developing calorie-reduced foods in which the calorie reduction would be accomplished through fat reduction and/or incorporation of fat-mimicking materials.

Fats and oils directly or indirectly affect all of the sensory aspects of foods, but profoundly affect texture. Because of the very diverse and complex role of fat in food systems, no simple classification system can totally cover its functionality or its contribution to sensory properties. Fat can act as a lubricant on the molecular level. It also is a modifier of properties of other ingredients. For example, without fat in baked goods, gluten and starch will not have the proper structure. Fat affects mouthfeel in products such as salad dressings and milk.

Brandt et al.[1] state that fat and moisture content are multidimensional and do not lend themselves to being scaled across an entire texture range as do the mechanical parameters of texture.

Prior to developing procedures for evaluating foods containing fats, work was undertaken on the evaluation of fats and oils themselves. Sensory knowledge and procedures for fats and oils evaluation would provide a background upon which to develop procedures for evaluating foods. Also, such a procedure would provide the mechanics for assessing "fat-likeness" of fat-mimicking material.

A. Fats and Oils

Several fats and oils were examined to develop a vocabulary for describing their mouthfeel/texture. Commercial emulsions, as well as starches and gums, were also included (see Table 6). These materials were selected because they represent:

1. Ingredients used in a wide variety of processed foods as well as in home food preparation
2. Fats such as butter and margarine, which are consumed in unaltered form, usually as spreads
3. A range of melting points covering those that are liquid at room temperature and those that are solid at room temperature

Table 6
REFERENCES FOR FATS/OILS EVALUATION

	Solids at room temperature			Liquids at room temperature	
Fats	**Water-in-oil emulsions**	**Other**		**Oils**	**Other**
Crisco®	Butter	Corn starch solutions (12 and 15%)		Olive	CMC gum solutions (2, 3.5, and 5%)
Hydrogenated soy-bean oil	Whipped butter			Wesson®	
Hydrogenated and fractionated coconut oil	Margarine			Peanut Safflower Soybean	
Lard				Mineral	
Cocoa butter				Castor Captex 300®	

Table 7
DESCRIPTIVE TERMS AND DEFINITIONS FOR FATS AND OILS

Term	Definition
Density	Perceived weight of sample on the tongue
Viscosity	Perceived thickness of sample
Wetness	Perceived moistness of sample surface
Slipperiness	Degree of lubrication perceived on lips
Spreadability	Degree to which sample spreads over the tongue
Melt impression	Thermal breakdown; degree to which sample perceptibly dissolves from a solid or semi-solid to a liquid
Adhesiveness	Force required to remove the material that adheres to the mouth
Resistance to tongue movement	Degree of force required to manipulate product

Corn starch gels and gum solutions were included since they are conventionally used as thickeners and may be thought of as fat-mimicking materials. Also, they would aid in vocabulary development by their sensory differentiation from fats and oils.

Applicable terms and definitions for describing fats and oils are shown in Table 7. The intrinsic sensory characteristics of fats and emulsions are slipperiness, melt impression, and mouthcoating.

Oils, like fats, are high in slipperiness and mouthcoating. However, the sensory characteristic of melt impression in which the fat seems to liquefy in the mouth is lacking in oils. The CMC gum solutions and corn starch gels lack slipperiness, melt impression, and mouthcoating. Table 8 shows the degree of each texture characteristic in these reference materials in terms of low, medium, and high, and also lists their order of perception.

B. Fat-Containing Foods

A wide variety of fat-containing foods was selected initially to orient the panel to a diversity of textures. The foods ranged from liquid to solid systems and covered a wide range of fat content (Table 9).

In liquid systems such as nonstabilized salad dressings, the presence of oil can be perceived directly. The sensory descriptors are similar to those used to describe oils, i.e., slippery and mouthcoating. Other liquids, such as light cream, are described as slippery, smooth, and mouthcoating.

In semisolid systems such as whipped toppings and cheesecakes, fat per se becomes less perceptible in the mouth. Rather, it affects texture/mouthfeel characteristics which can be

Table 8
TEXTURE CHARACTERISTICS OF FATS, OILS, GUMS, AND STARCHES

	Solids at room temperature			Liquids at room temperature	
Texture/mouthfeel characteristics	Fats	Water in oil emulsions	Corn starch gels	Oils	CMC gum solutions
Density	Medium	Medium	Medium	Low	Low
Viscosity	None	None	None	Medium	Medium
Wetness	Medium	Medium	High	None	None
Slipperiness	High	Medium	None	High	None
Spreadability	None	None	High	High	High
Melt impression	Medium	High	None	None	None
Adhesiveness	Medium	Medium	None	None	None
Resistance to tongue movement	Medium	Medium	Medium	Low	Low
Mouthcoating	High	Medium	None	High	None

Table 9
FAT CONTAINING FOOD REFERENCES

Products	Fat (%)	Descriptors
Salad dressing	60	Slippery/mouthcoating
Light cream	19	Smooth/slippery/ mouthcoating
Whipped topping	26	Dry/smooth/firm/spread
Vanilla cookies	16	Crumbly/uniform/dry
Enriched white bread	3	Soft/moisture-absorbing/ gummy
Croissants/Danish pastries		Crispy/flaky/chewy/moist
Frankfurters	27	Juicy/springy/firm/chewy
Cheesecake	36	Heavy/smooth/adhesive/ mouthcoating

more specifically described as smooth, filmy, cohesive, adhesive, etc. The presence and degree of these characteristics are product-specific since other components of a food, as well as its fat content, contribute to texture/mouthfeel.

Fat and oil are not discernible as such in solid foods such as bread and cookies. Solid foods of high fat content such as Danish pastries and croissants are described as crispy, flaky, chewy, and moist. Frankfurters are described as juicy, springy, firm, and chewy.

An additional set of products was selected for development of vocabulary and evaluation procedures. This set was limited to liquids and semisolids since these were the systems of interest to the research and development team. Table 10 lists these products in increasing order of fat content. Texture terms and definitions developed through evaluation of these reference products are shown in Table 11.

C. Creaminess of Foods

The term *creaminess* is frequently used by consumers in describing foods. Szczesniak and Skinner showed that in free association testing, product responses to "creaminess" included cream, puddings, ice cream, butter, potatoes, frostings, sauces, and cheesecakes.[22]

The fat content of these products varies from 1% for potatoes to 80% for butter. In the case of potatoes, the product association with creaminess was probably mashed potatoes,

Table 10
FAT CONTAINING FOODS

Product	Fat (%)
Brand X frozen dessert	0.8
Vanilla puddings	1.7
Yogurt (plain)	1.8
Milk	3.5
Ice milk	5.1
Evaporated milk	7.9
Imitation sour cream	10.6
Half and half cream	11.7
Ice cream	10—16
Sour cream	17.0
Light cream	19.3
Cool Whip®	25.7
Cream cheese	35.3
Heavy cream	37.6
Cheesecakes	35—38

Table 11
TEXTURE TERMS AND DEFINITIONS FOR LIQUID AND SEMISOLID REFERENCE

Term	Definition
Surface wetness	Degree of moisture on the surface
Temperature	Thermal effect on the tongue; sense of warmth/coldness
Heaviness	Weight of product as perceived when placed on tongue
Firmness	Amount of force required to compress partially or fully
Smoothness	Absence of discrete particles
Adhesiveness	Force required to remove the material that adheres to the mouth
Spreadability	Degree to which the product spreads over the tongue
2-Phase impression	Simultaneous occurrence of more than one textural parameter
Gumminess	Energy required to disintegrate a semisolid sample to a state ready for swallowing; cohesiveness
Moisture absorption	Degree to which the sample mixes with saliva
Rate of disappearance	Time required for breakdown (slow, moderate, fast)
Uniformity of disappearance	Degree to which the product remains uniform throughout the breakdown
Type of disappearance	Degree to which the product reduces from its original state to a thin liquid state
Mouthcoating	Type and degree of coating in the mouth after manipulation
Type of residual mouthfeel	Identify and intensify any residual feelings remaining in mouth

creaminess being associated with their soft, smooth texture. Puddings fit this example as well. Creaminess is probably automatically associated with foods in which cream is part of the product description, e.g., ice cream. Also, creaminess may simply be used as a way of indicating that a product or recipe contains cream as an ingredient, e.g., sauces.

Because "creaminess" is part of the everyday food vocabulary of consumers and food developers, and also because of its apparent multiplicity of meanings and associations, it was decided to incorporate this term into the texture profiling of liquid and semisolid systems to strengthen the communication bridge between the laboratory and the consumer.

Several commercial brands of each product type that was frequently associated with the term "creaminess" were selected for panel evaluation. Texture profiles were developed on all samples and the common terms were identified. The common terms which characterized the product array were

- Smoothness (absence of discrete particles)
- Mouthcoating (fatty/oil/filmy)
- Viscosity (thickness during breakdown)

It was not possible to quantify the contribution of each of these three parameters to creaminess. However, the lack of any one or more of them will result in a product score of very low or no creaminess.

Those brands which were premium quality and/or market leaders were assigned a score of 10 or higher on a 14-point scale where 0 = not creamy and 14 = extremely creamy.

The creaminess scale was incorporated into the ballot as the final score assigned to test samples during product evaluation. Prototypes, either full-fat or low-fat, could then be compared to the "gold standard" product for comparative creaminess. The advantage of this scale is that it provides a single rating which simplifies communication with the food researcher and also can be compared with consumer results while, at the same time, it can be tracked for the specific characteristic which contributed to the creaminess score.

D. Application to Cheesecakes

Cheesecakes were selected to evaluate the utility of the methodology. Four cheesecakes — one scratch recipe and three commercial products — were used for this purpose.

Six panel members from the pool who had undergone training for evaluating fats and fat-containing foods were selected to form the cheesecake panel. All panel members evaluated all four cheesecakes within a 3-week period. Randomized order of sample presentation was used and replicate evaluations were obtained.

Principal Components Analysis was used to analyze these data. Results of the analysis indicated that differences among these cheesecakes could be represented by two factor dimensions, factor 1 being high adhesiveness/low spreadability and factor 2 being high smoothness/low two-phase impression. The relationships among the cheesecakes in terms of these factors are shown in Figure 1.

The scratch recipe was very low on smoothness and moderately high on adhesiveness, while Brand B was the reverse. Brands A and C fell roughly between the others.

The creaminess ratings which were assigned to these cheese cakes are shown in Figure 1 as well. Creaminess rating increased with increased perception of smoothness and decreased perception of adhesiveness.

The high reliability of the panel data can be seen by the closeness of the replicate samples in the factor space.

The unreduced data in the form of the Texture Profiles are shown in Figures 2 to 6. Figure 2 depicts the profiles of the four cheesecakes; Figures 3 to 6 show replicated data for each cheesecake.

VIII. OVERVIEW

By virtue of its broad use in the food industry, the Texture Profile has proven to be a valuable addition to the battery of tests available for measuring the sensory quality of foods and beverages. R & D teams responsible for creating new or improving current food textures benefit from its detailed fingerprinting of individual texture characteristics. It is indispensable to the researcher in developing instrumentation for measuring food texture. The panels trained in the use of the Texture Profile form an effective communications bridge between the laboratory and the consumer in the ongoing dialogue of food texture measurement and acceptance.

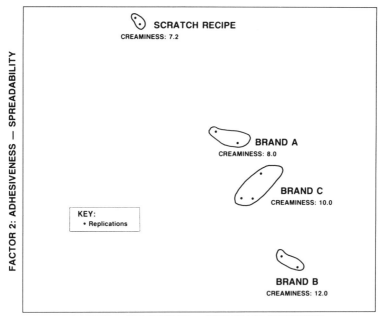

FIGURE 1.   Principal components analysis of the texture of cheesecakes.

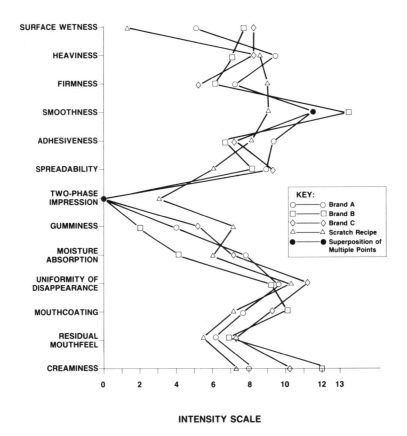

FIGURE 2.   Texture profiles of cheesecakes.

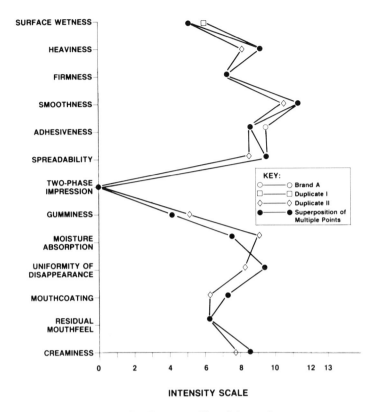

FIGURE 3. Texture profiles of cheesecakes.

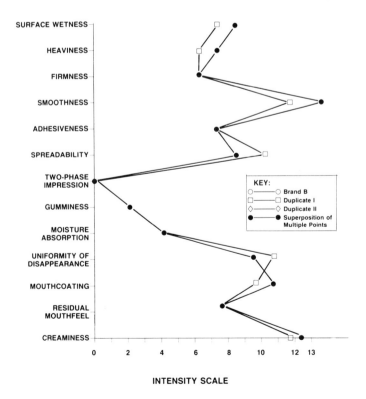

FIGURE 4. Texture profiles of cheesecakes.

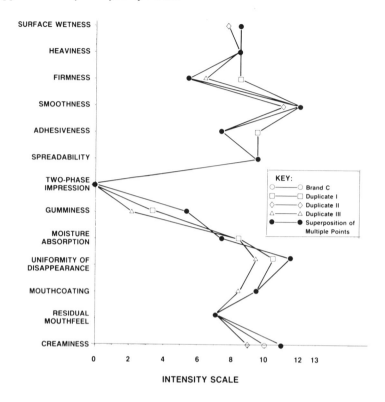

FIGURE 5. Texture profiles of cheesecakes.

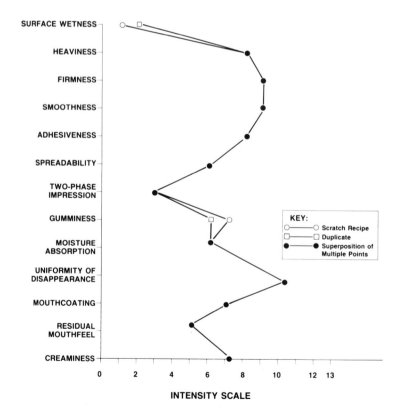

FIGURE 6. Texture profiles of cheesecakes.

REFERENCES

1. **Brandt, M. A., Skinner, E. Z., and Coleman, J. A.,** Texture profile method, *J. Food Sci.*, 4(28), 404, 1963.
2. **Szczesniak, A. S.,** General Foods texture profile revisited — ten years perspective, *J. Texture Stud.*, 1(6), 43, 1975.
3. **Bourne, M. C., Sandoval, A. M. R., Villabolos, M., and Buckle, T. S.,** Training a sensory texture profile panel and development of standard rating scales in Colombia, *J. Texture Stud.*, 1(6), 43, 1975.
4. **Civille, G. V. and Szczesniak, A. S.,** Guidlines to training a texture profile panel, *J. Texture Stud.*, 2(4), 204, 1973.
5. **Cardello, A. V., Maller, O., Kapsalis, J. G., Segars, R. A., Sawyer, F. M., Murphy, C., and Moskowitz, H. R.,** Perception of texture by trained and consumer panelists, *J. Food Sci.*, 47, 1186, 1982.
6. **Moskowitz, H. R., Kapsalis, J. G., Cardello, A. V., Fishken, D., Maller, O., and Segars, R. A.,** Determining relationships among objective expert and consumer measures of texture, *Food Technol.*, 10(33), 84, 1979.
7. **Moskowitz, H. R. and Kapsalis, J. G.,** The texture profile: its foundation and outlook, *J. Texture Stud.*, 1(6), 157, 1975.
8. **Arocha, P. M. and Toledo, R. T.,** Descriptors for texture profile analyses of frankfurter-type products from minced fish, *J. Food Sci.*, 3(47), 695, 1982.
9. **Pangborn, R. M. and Koyasako, A.,** Time-course of viscosity, sweetness and flavor in chocolate desserts, *J. Texture Stud.*, 2(12), 141, 1981.
10. **Ingate, M. R. and Christensen, C. M.,** Perceived textural dimensions of fruit-based beverages, *J. Texture Stud.*, 2(12), 121, 1981.
11. **Schwartz, N. O.,** Adaptation of the sensory texture profile method to skin care products, *J. Texture Stud.*, 1(6), 33, 1975.
12. **Civille, G. V. and Liska, I. H.,** Modifications and applications to foods of the General Foods sensory texture profile technique, *J. Texture Stud.*, 1(6), 19, 1975.
13. **Szczesniak, A. S., Loew, B. J., and Skinner, E. Z.,** Consumer texture profile technique, *J. Food Sci.*, 6(40), 1253, 1975.
14. **Moskowitz, H. R.,** Intensity scaling for product testing, *Product Testing and Sensory Evaluation of Foods,* Foods and Nutrition Press, Westport, Conn., 1983, 5.
15. **Stone, H. and Sidel, J. L.,** Components of measurement, *Sensory Evaluation Practices,* Academic Press, Orlando, Fla., 1985, 3.
16. **Liska, I. H., Marcucci, M., and Walchak, C. G.,** unpublished data, 1985.
17. **Cairncross, S. E. and Sjostrom, L. B.,** Flavor profiles — a new approach to flavor problems, *Food Technol.*, 4(8), 308, 1950.
18. **Stone, H. and Sidel, J. L.,** Descriptive Analysis, *Sensory Evaluation Practices,* Academic Press, Orlando, Fla., 1985, 6.
19. **Szczesniak, A. S., Brandt, M. A., and Friedman, H. H.,** Development of standard rating scales for mechanical parameters of texture and correlation between the objective and sensory methods of texture evaluation, *J. Food Sci.*, 4(28), 397, 1963.
20. **Powers, J. J.,** Using general statistical programs to evaluate sensory data, *Food Technol.*, 6(38), 74, 1984.
21. **Syarief, H., Hamann, D. D., Gresbrecht, F. G., Young, C. T., and Monroe, R. J.,** Interdependency and underlying dimensions of sensory textural characteristics of selected foods, Journal Series of North Carolina Agricultural Research Service, Raleigh, N.C.
22. **Szczesniak, A. S. and Skinner, E. Z.,** Meaning of texture words to the consumer, *J. Texture Stud.*, 3(4), 378, 1973.

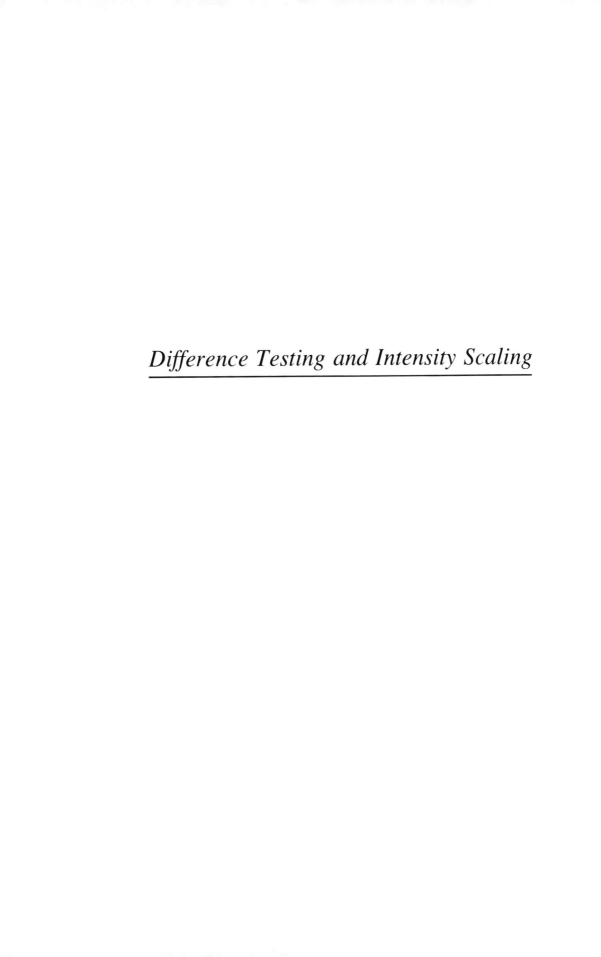

Difference Testing and Intensity Scaling

Chapter 6

DIFFERENCE TESTING: PROCEDURES AND PANELISTS

Alan S. Kornheiser

TABLE OF CONTENTS

I.   Introduction 112

II.   Identification Testing 112
       A.   Types of Tests... 112
       B.   Test Design 113
       C.   Data Analysis 114

III.   Signal Detection 115
       A.   Introduction 115
       B.   Analysis... 117
       C.   Example... 118
       D.   Test Procedure 118
       E.   Informal Testing.. 119

IV.   Comparison Testing.. 119
       A.   Types of Tests.. 119
       B.   Paired Comparison Hedonic Tests...................................... 119
       C.   Multiple Comparison Hedonic Tests.................................... 120
       D.   Paired Comparison Attribute Tests 120
       E.   Multiple Comparison Attribute Tests 121

V.   Panelists 121

VI.   Discussion 122

I. INTRODUCTION

The most basic form of product testing is difference testing — are two things the same or do they differ in some way? This is the screen through which all forms of product development and change must initially pass: a product "improvement" that cannot be detected is not an improvement; a manufacturing change that can be detected has passed, quite unconsciously, from the realm of manufacturing into the realm of product development; three new products that cannot be distinguished are, in fact, only one new product.

A surprisingly large amount of time, money, and effort ride on the results of difference tests. If one is confident of his test results, major savings can result and significant risks can be avoided or, even better, accurately evaluated. Given the importance of difference testing, manufacturers pay it surprisingly little attention and do less than they should — although when the author was at Pepsi-Cola, his lab alone conducted approximately 500 such tests each year. This paper summarizes what he knows about difference testing at Pepsi: how it was done, who it was done with, and what should have been done. It is the result of 18 months of research and a great deal of experience, all aimed at answering one question: how should difference testing be done?

Although all difference testing aims at answering the one basic question of whether products differ in a meaningful way, there are three somewhat different ways to approach the question. We may think of these approaches as identification, signal detection, and attribute. In an identification paradigm, a panelist states which of three products is the most different or which of two products is more like a third. This is the most common type of formal procedure; it is used under such names as triangle and duo/trio test. In a signal detection test, a panelist states whether two products are the same or different. This is the most common informal test and is widely used for quality control. Finally, a panelist may state which of two products has more of some carefully defined quality; when that quality is hedonic, (i.e., overall liking), this becomes a paired-preference test. Each of these tests has an appropriate place in difference testing, and it is an unusual researcher indeed who cannot find a use for each of these tests at one time or another. The discussion that follows focuses on how, when, and why each of these types of tests should be conducted. It begins with the most well-known type of testing — identification testing.

II. IDENTIFICATION TESTING

A. Types of Tests

The best-known type of identification testing is the triangle test. Two somewhat less common tests are the duo/trio and dual standard tests. Other possibilities exist, but they are rarely seen. All are inherently nonparametric; that is, they can tell if two products differ, but they cannot determine by how much, at least not directly. None is usable in a multiple-comparison format without significant difficulty. It is best to use these tests when a very simple answer is needed to the very simple question, "Are these two things different?"

In a triangle test, a respondent is given three products. Two of these are the same, one is "different". The respondent attempts to identify the "different" product; if he cannot, he is asked to guess. If he — or the panel as a whole — guesses right significantly more often than one third of the time, one concludes that a detectable difference exists.

In a duo/trio test, one product is identified as a reference. The respondent is also given two more products, one the same as the reference and the other "different". He is asked to identify the product that is the same as (or most similar to) the reference. If he — or the panel as a whole — guesses right significantly more often than half of the time, one concludes again that a detectable difference exists.

In a dual standard test, two references are provided, one of Product A and one of Product

B. The respondent is also given two more products, again one of Product A and one of Product B, and asked to match the two identified products with the two unidentified products. As random is still one half in this case, a score of significantly greater than 50% correct implies a detectable difference between Products A and B.

One frequently hears that the triangle test is "more sensitive" than the duo/trio test because its chance probability is lower. In a sense this is certainly true — the odds of a respondent getting three triangle tests right in a row just by chance is only $1/3 \times 1/3 \times 1/3 = 1/27 = 3.7\%$. The odds of getting three duo/trio tests correct by chance are $1/2 \times 1/2 \times 1/2 = 1/8 = 12.5\%$. Obviously, if there is an easily detected difference, fewer trials are needed with a triangle test than with a duo/trio test, and the triangle test is more sensitive. However, a triangle test requires three comparisons: a respondent compares, say, A vs. B, B vs. C, and finally A vs. C to determine which pair is most dissimilar. In a duo/trio test, only two comparisons are needed: reference vs. A, reference vs. B. In a test with minimal fatigue factors, such as evaluating color differences or texture differences, this difference is irrelevant. In those cases where complex and difficult-to-describe differences exist, coupled with strong tastes or aromas that induce fatigue, this difference can be most important.

In a series of carefully controlled taste tests using commercial soft drinks, triangle tests were found to be significantly *less* sensitive than duo/trio tests. A panel of 36 respondents was consistently able to detect differences using duo/trio tests which it could not detect using triangle tests. For any kind of testing in which fatigue could conceivably be a factor, duo/trio tests are clearly to be preferred to triangle tests. This is not to say triangle tests are not effective. For simple differences, especially where fatigue is not a factor (as, for example, with the products used in Gridgeman's well-known study [*J. Food Technol.*, 9, 148, 1955]), triangle tests are just as sensitive as duo/trio tests, and their statistical advantage makes them more efficient. Only where fatigue is a factor do duo/trio tests excel.

There is no theoretical advantage to dual standard tests. They require four comparisons — reference 1 vs. A, reference 1 vs. B, reference 2 vs. A, reference 2 vs. B — just as a triangle test does, but have a chance detection rate of 50%, just as a duo/trio test does. However, panelists tend to like them. Although the extra reference does not, in fact, supply any extra information, panelists believe it does and feel more confident of their test accuracy. Dual standard tests are therefore appropriate where nonexpert panels are used and morale is an important consideration.

B. Test Design

It should go without saying that nothing in the test should provide any extraneous cues to the panel. The order of presentation of products in difference tests must be randomized. One cannot, for example, always put the "different" product in the middle for a triangle test and expect clean results. Symbols used to identify products should be randomized — three digit codes are common, but almost anything will work as long as it is systematically varied. The list of things to be controlled is almost endless, but it is all merely a matter of common sense.

A more interesting question is whether the odd product in a triangle test or the reference in a duo/trio test should be rotated. If, for example, one wishes to change sugar suppliers for a soft drink and is comparing soda made with the old and the new sugar, what is the appropriate reference product in a duo/trio test? Should it be the product made with the standard sugar or should the two be alternated? If a triangle test is being used, should the "different" product always be the new product, or again should the researcher alternate? Reasonable men may differ, which may be a good indication that it really does not matter very much.

Theoretically, one wishes to avoid any possibility of a learning curve in a difference test, which implies that the reference should be rotated. What, however, if the test is not a psychophysical threshold test in an academic laboratory, but an attempt to minimize the risk of losing market share while saving the company money? In that case, one is not trying to find the average difference threshold; he is trying to minimize the risk of anyone drinking our soda and thinking it tastes different. Under those circumstances, anything that increases test sensitivity is a good thing, and the reference should not be rotated. This brings up a good general rule: if one wishes to determine a difference threshold — and that is what one usually wants in a psychophysical examination — the reference should by all means be rotated. If, on the other hand, one wishes to determine whether a difference is too small to be detected — and that is what one normally wants in industry — the present product should be used as the reference, not the test product.

C. Data Analysis

The nature of this kind of difference test requires that testing be done with a fairly large panel. How big is big enough? It should be big enough so that even if a difference is too small for most of your panel to detect, a significant difference will be detected; but small enough not to be awkward and unnecessarily expensive. Experience has shown that a panel of about 24 is close to optimum. A larger panel does not greatly increase sensitivity, while a smaller panel runs the risk of missing a difference. Your panel may be 24 different people or 12 people who evaluate each set of products twice; fewer than 12 different people is risky.

You can further minimize your risks of rating different products "not different" by establishing a retest rule: products rated different at a 95% confidence level are reported as different; products rated different at less than a 90% confidence level are reported as not different; products rated different at 90 to 95% confidence levels are reported as marginal and retested.

A certain amount of common sense is also helpful. If a series of related tests are being run and one variant is rated different at the 94% confidence level and another is rated different at the 96% confidence level, it should be fairly obvious that these two products differ from the control by about the same amount. Either that should be said, or, better yet, both should be retested. In any event, one should not blindly follow some arbitrary cutoff and say one is "different" and the other is not. Statistics are only a servant, and as the White Knight told Alice (in a somewhat different context), the question is, "Who will be the master?"

Although it seems too obvious to need mentioning, all analysis is one-tailed. Results below chance level, no matter how far below chance level they are, cannot be significant. They can, of course, indicate that something is being done wrong. If only 10% of the panel is correct on a duo/trio test, one product may have been misidentified, or the panel may have misunderstood the instructions.

When looking to see whether the differences that have been found are significant, most researchers simply keep a table convenient. Technically, distributions are binomial (with $p = 1/3$ and $q = 2/3$ for triangle tests and $p = q = 1/2$ for duo/trio and dual standard tests); however, everyone assumes his sample is sufficiently large for a normal approximation and uses a table of t scores or z scores. If the sample is too small (below ten or so), this is not the world's greatest approximation and the actual binomial calculation probably should be used.

Possibly the single most common error in this type of testing is confusing the percentage of correct answers with the percentage of the test population capable of making a discrimination. If 75% of the panelists correctly identify the reference in a duo/trio test, only about 50% of the panelists can be assumed to have correctly detected the difference; of the remaining 50%, half (i.e., 50% of 50% = 25%) guessed right and half guessed wrong. Combining the 25% "luckies" with the 50% "accurates" yields the 75% who responded correctly. For duo/trio tests, assuming above-random detection, this formula holds:

$$\text{Percentage Actually Detecting} = (2 \times \text{Percentage Correct}) - 100\%$$

For triangle tests, this formula holds:

$$\text{Percentage Actually Detecting} = (3/2 \times \text{Percentage Correct}) - 50\%$$

Such simple arithmetic can have significant consequences. In a good-sized test, finding that as few as 16% of your test population was correct on two successive triangle tests can be highly significant. However, the percentage of the population that actually detected the difference twice may be no more than 6% of the population. Do you really want to make a change that no more than 6% of your market can reliably detect? Never lose track of the distinction between a "statistically significant" difference and a large difference.

III. SIGNAL DETECTION

A. Introduction

All true signal detection testing works the same way. A panelist is given two products and is asked if they are the same or different. Sometimes the products will be the same; other times the tester will not know if they are the same. (Those pairs which the tester knows are different are the control trials.) When the panelist calls the same products "different", this is called a false alarm. Those times he calls unlike products "different" are "hits". If, for a given pair of products, the percentage of hits is significantly greater than the percentage of false alarms, the two products are said to be detectably different.

This system has several advantages: (1) it is very sensitive, especially for products in which sensory fatigue is a factor; (2) it is very natural; this is how people think about testing, no matter how you train them with duo/trio or triangle tests; and (3) this system can be used quantitatively with very small samples and qualitatively with truly tiny samples. The reason one sees signal detection so seldom is that it is very difficult to administer properly.

Although one might argue that all testing is really signal detection testing, this is the classic problem of discriminating between real events and false alarms. One of the classic examples is a radar operator watching his screen. The screen is noisy — full of snow, chaff, etc. Every so often, an airplane will be on the screen. The operator's problem is to accurately identify each airplane ("detection") without incorrectly identifying some piece of noise as an airplane ("false alarm"). The task is inherently impossible; if he only reports an airplane when he is absolutely certain it is not noise, he will miss most of them (low false alarm rate but low detection rate). If he reports an airplane whenever he sees something that may be a plane, he will get most of them but will also report many airplanes that are not really there (high detection rate but high false alarm rate).

In signal detection theory, one studies the relationship between the false alarm rate and the detection rate. For any given set of signal and noise (i.e., strength of a plane's radar echo and the amount of "snow" on the radar), many systematic pairs of detection and false alarm rate are possible. The detection rate and the false alarm rate are increased by rewarding correct detection and not punishing false alarms. Detection and false alarms are decreased by reducing the reward for detection and increasing the punishment for false alarms. Eventually, an optimal level of false alarm and detection can be found that fits the needs of a particular study.

Sensory panels are in the same position as the radar operator. In a typical study, a panelist is given two products and asked if they are the same or different. Two of *anything* are always slightly different; the act of sensing something changes that sense — no man can step twice into the same river. Thus, the natural tendency of an untrained panelist is to call almost everything "different". Indeed, the natural false alarm rate of an untrained panelist hovers around 50%. However, these rates are easily manipulated. The easiest manipulation is simple feedback: if panelists are told they are wrong whenever they have a false alarm, the rate drops. If they are told that missing a difference is much less important than keeping their false alarm rate down, the false alarm rate decreases even more. More complicated, but more fun, is to actually establish a series of rewards and punishments — say, $.25 for a correct detection but a fine of $.50 for a false alarm.

Now the problems can be seen. Two responses must be measured, and neither of them is a simple function of the products being tested. In a way, then, signal detection is at least twice as complex as simple triangle testing. Actually, as will be seen, the situation is even more complex, but there are times when its advantages make this technique a very desirable test option.

In this type of testing, a respondent is required to state whether two products are the "same" or "different". Because the overall false alarm rate must be determined, some of the trials must indeed be two of the same product. However, it must be ensured that respondents are not becoming too conservative and saying "same" to all pairs; some trials must be included in which products known to be different are presented. A signal detection test is, therefore, a combination of three different tests — true trials, false alarm trials, and known-different trials. Typically, about two thirds of all trials are true trials, one sixth are false alarms, and one sixth are known-different trials. These ratios may change if a panel requires training or retraining.

In any given test, each panelist is given two products and asked if they are the "same" or "different". (Of course, panelists never know if a trial is a true, false alarm, or known-different trial.) The percentage of respondents calling the product "different" is then calculated. At this point, one of three calculations is made, depending on the type of trial.

If this was a false alarm trial, the data are added to a running false alarm tally that is kept for each respondent and for the panel as a whole. Typically, the rate will average about 10 to 25%; that is, 10 to 25% of the time a respondent will call two identical products "different". If a respondent's rate begins to increase — that is, if he begins to call many of the same products "different" — feedback and retraining are needed. The running tally of false alarm rates should reach back about 1 month or 15 trials, whichever is less; false alarm data should be kept as current as possible. That a respondent had a high false alarm rate 2 months ago has little bearing on his current false alarm rate.

In a known-different trial, the data are added to the running "difference" tally. This should average above 75%; if a panelist's score drops below this level, retraining is usually indicated.

False alarm rates and difference rates correlate highly — a panelist with a high false alarm rate and a low difference rate is rare and undesirable; if retraining cannot cure the problem, the panelist should be dropped.

If the trial was a true trial, the data are then analyzed to determine whether the panel called the product different at a level significantly higher than the false alarm rate.

The recordkeeping can be complicated and tedious. Figure 1 shows one set of records; ideally, the entire process should be computerized and fully automated, a not-very-complicated process.

Duo/trio and similar tests do not require such elaborate recordkeeping. Their baseline chance levels are inherent in their designs — the "false alarm" level for a triangle test is 1/3. Why, then, bother with this procedure? Done correctly, it can be incredibly sensitive with very small samples.

ACCURACY RECORD

Panelist Name	FALSE ALARM SCORES					KNOWN-DIFFERENT SCORES			
	10/12	10/14	10/18	10/19	10/21	10/13	10/15	10/16	10/22
A. ABRAMS	$+(\frac{3}{12})$	$-(\frac{3}{13})$	$-(\frac{3}{14})$	$0(\frac{3}{14})$	$-(\frac{3}{15})$	$-(\frac{9}{12})$	$+(\frac{10}{13})$	$+(\frac{11}{14})$	$+(\frac{12}{15})$
B. BROOKS	$+(\frac{3}{12})$	$+(\frac{4}{13})$	$-(\frac{4}{14})$	$+(\frac{5}{15})$	$+(\frac{6}{16})$	$+(\frac{11}{12})$	$+(\frac{12}{13})$	$+(\frac{13}{14})$	$+(\frac{14}{15})$
C. CARNEGIE	$-(\frac{3}{12})$	$-(\frac{3}{13})$	$+(\frac{4}{14})$	$+(\frac{5}{15})$	$-(\frac{5}{16})$	$+(\frac{11}{13})$	$+(\frac{12}{14})$	$+(\frac{13}{15})$	$+(\frac{14}{16})$
D. DUCK	$-(\frac{1}{13})$	$-(\frac{1}{14})$	$-(\frac{1}{15})$	$-(\frac{1}{16})$	$-(\frac{1}{17})$	$-(\frac{8}{12})$	$+(\frac{9}{13})$	$+(\frac{10}{14})$	$-(\frac{10}{15})$
E. ELMSFORD	$-(\frac{1}{12})$	$+(\frac{2}{13})$	$-(\frac{2}{14})$	$-(\frac{2}{15})$	$-(\frac{2}{16})$	$+(\frac{11}{12})$	$+(\frac{12}{13})$	$+(\frac{13}{14})$	$+(\frac{14}{15})$
F. FISH	$-(\frac{0}{11})$	$-(\frac{0}{12})$	$-(\frac{0}{13})$	$0(\frac{0}{13})$	$-(\frac{0}{14})$	$0(\frac{9}{11})$	$-(\frac{9}{12})$	$+(\frac{10}{13})$	$+(\frac{11}{14})$
G. GOOSE	$0(\frac{1}{11})$	$-(\frac{1}{12})$	$0(\frac{1}{12})$	$-(\frac{1}{13})$	$-(\frac{1}{14})$	$+(\frac{10}{12})$	$+(\frac{11}{13})$	$+(\frac{12}{14})$	$0(\frac{12}{14})$
H. HIERO	$+(\frac{1}{13})$	$-(\frac{1}{14})$	$-(\frac{1}{15})$	$-(\frac{1}{16})$	$-(\frac{1}{17})$	$+(\frac{12}{13})$	$+(\frac{13}{14})$	$+(\frac{14}{15})$	$+(\frac{15}{16})$
I. INGLES	$+(\frac{1}{12})$	$-(\frac{1}{13})$	$-(\frac{1}{14})$	$-(\frac{1}{15})$	$-(\frac{1}{16})$	$-(\frac{6}{11})$	$+(\frac{7}{12})$	$-(\frac{7}{13})$	$+(\frac{8}{14})$
J. JACKSON	$-(\frac{2}{11})$	$-(\frac{2}{12})$	$-(\frac{2}{13})$	$-(\frac{2}{14})$	$+(\frac{3}{15})$	$+(\frac{8}{12})$	$0(\frac{8}{12})$	$+(\frac{9}{13})$	$+(\frac{10}{14})$
K. KORNHEISER	$-(\frac{0}{13})$	$-(\frac{0}{14})$	$-(\frac{0}{15})$	$-(\frac{0}{16})$	$-(\frac{0}{17})$	$+(\frac{13}{13})$	$+(\frac{14}{14})$	$+(\frac{15}{15})$	$+(\frac{16}{16})$
L. LEWIS	$-(\frac{1}{9})$	$0(\frac{1}{9})$	$-(\frac{1}{10})$	$+(\frac{2}{11})$	$0(\frac{2}{12})$	$+(\frac{8}{9})$	$+(\frac{9}{10})$	$0(\frac{9}{10})$	$0(\frac{9}{10})$
Panel average:	$\frac{4}{12}(\frac{17}{141})$	$\frac{2}{11}(\frac{19}{152})$	$\frac{1}{11}(\frac{20}{83})$	$\frac{3}{10}(\frac{23}{173})$	$\frac{2}{11}(\frac{25}{184})$	$\frac{8}{11}(\frac{116}{182})$	$\frac{10}{11}(\frac{126}{153})$	$\frac{10}{11}(\frac{136}{164})$	$\frac{9}{10}(\frac{145}{174})$

Key: + Products called "different"

   - Products called "same"

   0 Panelist did not take test

Procedure: Tally all "+" marks within the brackets as numerator.
Tally all the "+" and "-" marks as denominators. False
alarm rates should remain below 1/5; known-different
detection rates should remain above 4/5.
Carry averages across to next page only twice: averages
should be means of last 13 to 18 trials.

FIGURE 1.   Sample record for a signal detection sensory panel.

B. Analysis

For a product to be significantly different from a control, it must be called "different" at a significantly higher rate than the false alarm rate. That is, in the following equation, z must be significant:

$$z = \frac{D - FA}{\sigma}$$

where FA = the overall panel false alarm rate and D = the difference rate (i.e., the percentage of panelists calling the two products "different").

The formula for σ is essentially the formula for the variance of the difference between two normal distributions:

$$\sigma^2 = \frac{\hat{p}(1 - \hat{p})}{N_{FA} \, N_D(N_{FA} + N_D)}$$

where $\hat{p}$ is the expected overall difference rate:

$$\hat{p} = \frac{(N_{FA} \times FA) + (N_D \times D)}{N_{FA} + N_D}$$

where N_{FA} = the total number of panelists who have generated the false-alarm rate and N_D = the total number of panelists who have generated the difference rate.

The z is evaluated on a one-tailed normal distribution, since any negative z score inherently implies a nonsignificant difference. Should z be very large and negative, one should consider checking the panel.

This all looks much more difficult than it is. Once a panel has been in operation for a while, N_{FA} will be much larger than N_D and the equation for σ will simplify to:

$$\sigma^2 \cong \frac{FA(1 - FA)}{N_D}$$

Since FA should average about 0.20 in a well-trained panel, the equation for z will simplify to:

$$z = \frac{5(N_D)^{1/2}}{2} (D - FA)$$

C. Example

Assume a panel of 12, just starting out, has had only 4 false alarm trials. In those four trials, a total of seven panelists said the products were different. Thus:

$$N_{FA} = 48 \qquad FA = 7/48 = 14.6\%$$

On a difference trial, seven panelists say the two products are different. Thus:

$$N_D = 12 \qquad D = 7/12 = 58.3\%$$

Inserting these numbers into the equations yields:

$$D - FA = 0.437$$
$$\hat{p} = 0.233$$
$$\sigma = 0.137$$
$$z = 3.19$$

significant at the $p < 0.01$ level, even though a panel of only 12 was used and only 7 detected a difference. It is this sensitivity that makes up for all of the other complexities.

D. Test Procedure

Test procedures in signal detection are simple and straightforward, similar to those in other forms of testing. Respondents are given two products, neither coded in any way, and asked only to indicate if they are the same or different. They are not asked "how different" nor are they asked how confident they are that the products are different. Neither of these two additions has been found to increase accuracy, and they sometimes actually decrease it by confusing the panelists.

Normal control procedures should be followed. Order of presentation should be rotated so that half of the panelists evaluate one product first and half evaluate the other one first. Obviously, panelists should not be allowed to observe the products being prepared.

Tests should be scheduled regularly. About one test in six should be a false alarm trial and about one in six a control trial with known-different products. Should there be a lull in product testing these rates can be increased; should there be a temporary flurry of tests they can be decreased, but this should be done most carefully — false alarm and known-different trials are vital to calibration and analysis and cannot be easily dispensed with.

E. Informal Testing

Most quality control (QC) testing uses an informal version of this test. Products are selected randomly and a few "experts" judge whether or not they are different from what they should be. There are no controls — neither a known-correct control for comparison nor known-different controls to keep the QC inspectors on their toes.

The formal procedure described above is the superior method. However, sometimes it cannot be used due to time, budget, or personnel constraints, but this should not allow a totally uncontrolled test to stay in place. Even under the most informal circumstances, QC judges should be given controls; some percentage of all the products they evaluate should be defective, some should be known to be acceptable. Doing so greatly improves the believability of informal QC testing at remarkably little cost. While still not the best procedure, this method is better than many others.

IV. COMPARISON TESTING

A. Types of Tests

Comparison tests are usually divided into attribute and hedonic tests — that is, tests which determine whether a product has more of some attribute than another and those which determine whether a product is liked more than another. Such tests also can be divided into paired and multiple comparisons; in multiple comparison tests one rank orders three or more products according to some attribute or according to how much the products are liked.

This is the most common type of test, and for good reason — it requires minimal training and is easily understood. Under appropriate controls, it is also the most sensitive test possible. Inappropriately controlled, it is trouble waiting to happen.

B. Paired Comparison Hedonic Tests

This is the most common test there is, familiar to many, for example, as TV's Pepsi® Challenge. (The research backing up those TV claims, as it happens, usually involved home-use tests and was more complex than appeared on TV.) A panelist evaluates one product, then the other, and states which is preferred. The interval between evaluating the two products may vary, and for taste and similar tests the panelist may rinse his mouth or otherwise refresh herself between products.

Sometimes a panelist may be asked to rate the first product's acceptability — plus perhaps some other attribute — before evaluating the second product. The test then becomes a "protomonadic" test and is usually much less sensitive to differences than a true paired-comparison test. Sensitivity increases as the interval between the evaluations of the two products decreases.

Obviously, the order of presentation must be both fully balanced and randomized; order-presentation bias is often very strong. It is usually best not to code the samples for the panelists; they are simply allowed to respond "first" or "second" product. That is how they will think of the products anyway, whether or not they are coded.

It is vitally important to remember that finding no difference in preference does not imply that there is no difference in the products, nor does it truly imply equally desirable products — half of the respondents may hate one product while the other half very mildly prefer it.

Hedonic scales of various kinds provide better data. Paired preference tests are not really difference tests at all. However, they are frequently treated as such, so be wary.

Analysis is simple, as this is a simple binomial problem with chance being 50% of all panelists preferring one product and 50% preferring the other. Those who cannot state a preference cause a problem; the author prefers to both calculate preferences among those who state a preference and to divide those "no preference" panelists equally and recalculate the preferences. The first result provides what the author believes is the best measure of whether a detectable difference exists; the second provides the best measure of whether an important difference exists. For very small samples, use a table or calculate the results from the binomial formula; for samples of 20 or larger, the Poisson approximation is fine, using a two-tailed z score with:

$$z = \frac{\text{number preferring} - (\text{total sample})/2}{(\text{total sample})^{1/2}/2}$$

The "total sample" and the "number preferring" can either include the "no preference" respondents or not, as the researcher prefers.

C. Multiple Comparison Hedonic Tests

In this rather uncommon test, respondents are given three or more products and asked to rank them according to acceptability. As this is inherently a more difficult task than a paired test, one rarely sees it. Properly done, it can be statistically very sensitive, but there is rarely a need for it; it is usually better to break comparisons into pairs.

Obviously, order bias can be a problem in this type of test, and fully balancing the orders may be impossible for multiple products and a limited panel. Products must be labeled in this case, and that can also induce bias. The math gets tricky quickly; it is best to check with a statistical expert of some type before embarking on such testing.

Also, one must be wary of how these data are analyzed. If one looks at individual pairs within the multiple rankings, he cannot use simple paired comparison statistics; having inherently generated many pairs by the test, he must take that into account when calculating significance levels. The best procedure, especially if one has a hypothesis going into the test, is to analyze the data *in toto,* using one of several nonparametric statistical techniques. Again, an expert should be consulted before starting.

D. Paired Comparison Attribute Tests

Paired comparison attribute testing is inherently the most sensitive difference procedure available. By allowing panelists to focus on only one product characteristic — be it sweetness, color, texture, or cleaning ability — this method greatly heightens sensitivity. In fact, this method may be too sensitive in some cases — incredibly small differences, too small to affect liking in real use, can easily be detected. A certain amount of common sense can be helpful here.

Testing procedure is similar to that of paired comparison hedonic testing. Panelists are given two products and asked which has "more" of a given attribute. Order of presentation should be carefully balanced. Those who cannot find any difference can either be excluded from the analysis or have their results randomly allocated to the two test products, as in the case of hedonic testing. Analysis again assumes a simple binomial distribution, with chance being 50%, and a Poisson approximation as sample size becomes sufficient.

The problem with paired comparison attribute testing is that the attribute must first be understood. This is less trivial than one might suspect. Suppose the filter type was changed on a cigarette and a naive panel innocently asked to tell if it was "stronger" than a control. In an actual test of this kind, more than 12 different definitions of "strong" were assumed by the panel, including "more menthol" for a nonmentholated cigarette. More prosaically, it is well known that making a product sweeter will increase a chocolate's "chocolate flavor" and a cola's "cola flavor", while increasing the amount of chocolate or cola has no effect at all. Also, a panel asked to consider one attribute, even if it fully understands that attribute, cannot tell you about anything else. If one expects that a change will make a product smoother and it does something else, asking about "smoothness" will do little good. The moral is that paired comparison attribute testing should be used only when (1) the researcher is certain he is asking about the right attribute, (2) he is certain the panel understands the terminology being used, and (3) the researcher is not concerned about using an unrealistically sensitive test.

If the researcher has the slightest doubt as to whether he is looking at the correct attribute or whether the panel understands the terms being used, he should try some exploratory research first. A few panelists are given the products involved and asked how they would characterize the differences. If necessary, the panel should be trained about the attribute involved, using carefully controlled samples. Only after the researcher is comfortable that the panel fully understands the attribute and that the attribute is appropriate should he begin.

Analysis for paired comparison attributes is identical to that for paired comparison hedonics, with one exception: an initial hypothesis usually has been made as to which product has more of an attribute than does the other. In this case, testing will be one-tailed. Without such a hypothesis, testing will be two-tailed.

It usually is not a good idea to ask panelists more than the simple "Which has more?" question. Having determined that one product does "have more" of an attribute than another, rating scales can be used to quantify the data. Trying to combine these two tasks risks loss of sensitivity.

Looking at several attributes simultaneously is possible, but again loss of sensitivity is possible. There is, however, no reason why panelists cannot make several sequential comparisons, first looking for differences in attribute 1, then in attribute 2, and so on. However, when looking at multiple comparisons, the math again can become tricky. If the attributes are truly independent, thought should be given to adjusting the statistics to take the multiple comparisons into account.

E. Multiple Comparison Attribute Tests

Most of what has been discussed in the previous two sections holds for multiple comparison attribute tests. However, there is one context in which multiple comparison attribute tests are especially useful — as dose/response curves. When the issue involved is whether or not some substance has a specific effect or how much of a substance is needed before an effect can be detected, asking respondents to rank order products by attributes can be very effective.

Using the precautions previously described, respondents can be asked to rank products in terms of any given attribute and the results analyzed nonparametrically. (The math is not very difficult, but it is beyond the scope of this chapter.) Because the chance probability of ranking four products correctly, for example, is only $1/4 \times 1/3 \times 1/2 = 4.2\%$, even a very small sample can quickly reveal if there is a significant effect being measured. However, bear in mind that this is an inherently difficult task, especially if it involves taste or smell (which fatigue easily), and sensitivity will be less than in the case of a simpler paired comparison test.

V. PANELISTS

So far, there has been little mention about the panelists. Most difference testing uses moderately trained panelists — that is, respondents who are familiar with the researcher's procedures and terms but are not particularly trained or screened otherwise. Such panelists are usually superior to any expert panel or consumer panel for a number of reasons.

Consumer panels — that is, panels composed of randomly selected respondents in the field — are rarely well motivated. Their sensitivity to rather gross differences known to be well within the range of normal human sensitivity is frequently abysmally low. This does not reflect any inherent physical problems; rather, it seems likely that such panelists simply do not care, do not pay attention to instructions, or respond randomly out of boredom or general ennui. In carefully controlled tests, such randomly recruited panelists were shown to be markedly less sensitive to even easily detectable differences than were any other type of panelist.

Expert panels, composed of carefully trained and selected respondents, are also not well suited to difference testing. There is nothing wrong with their sensitivity or understanding of the testing, but they frequently become bored, attempt to second-guess the experimenter, resent instructions and correction, and are generally hard to manage. They are also an expensive resource. These problems are not terribly serious and would be acceptable if such panelists offered additional benefits. As it happens, they do not.

Pepsi's research has shown that trained, screened, expert panels are no more sensitive in difference testing than are much less expert panels. This is less surprising than one might think. Human sensitivity does not vary overwhelmingly; once one has removed the insensitive, there are few ultrasensitives. An expert is not so much someone who feels, tastes, or smells with greater sensitivity as he is someone who *understands* what is being seen, smelled, or tasted. A properly designed difference test obviates the need for this skill. In place of expert panels, panels of ordinary people with a small amount of screening and training and a healthy degree of motivation are recommended.

Panels may be recruited from outside a business or from within and may be paid cash, prizes, or nothing. These differences do not seem important. However, several things are important:

Motivation — Panels must want to do their job and want to do it accurately. Regular report cards describing each panelist's accuracy are helpful. High spirits are desirable, but horseplay is not. It must be possible to discipline those who get out of line.

Screening — About 10 to 30% of all prospective panelists, depending primarily on age, will have great trouble with the discriminations needed. Time, trouble, and effort can be saved by eliminating these prospects early. A set of screening tests using moderately easy differences should be passed by any prospective panelist before that panelist is used.

Training — Without overdoing it, a few hours of training to familiarize panelists with terminology, procedures, and what to expect should be a requirement. Panelists should not be patronized but should not be expected to know anything about testing until they have been shown it at least twice.

All of these conditions will be easier to meet if an adequate pool of panelists is available to draw on. To be resourceful in recruiting, the researcher should go outside his department or even his business. The more people available, the easier it is to have good people and to avoid overusing a few volunteers.

VI. DISCUSSION

The purpose of difference testing is to establish whether or not a detectable difference exists. Difference testing cannot show whether the difference is important. This is nothing to be worried about; there are other techniques to answer such questions.

However, there is a related question which constantly comes up that can and should be answered — how realistic is the difference measurement? After all, laboratory testing using trained and screened panelists is not the same as real world experience. Given that a difference is detectable — why should one care? When should one care?

There are two main questions here:

1. Just because a panel can detect a difference, can "real people" detect it?
2. How many "real people" can detect this difference and what will they do about it?

These are valid questions. After all, no more than half the population could ever tell the difference between Coke® and Pepsi® even before Coke® changed its formula. Fortunately, the questions have answers.

Basically, the answer to the first question is "yes". If a panel of 12 or 24 people can detect a difference easily enough for the result to be statistically significant, many people out in the real world will detect it, too — not everyone, not even a majority, but enough. Case histories are available for the unbelievers; Schlitz® beer is a good place to start.

The second question is harder to answer. People will put up with an awful lot without complaining, or even noticing. All that can be said for sure is that a much lower percentage of consumers will notice, as compared to your panelists. How much lower? Difference testing cannot answer that. Fortunately, there are other chapters in this book.

Chapter 7

ASSESSOR SELECTION: PROCEDURES AND RESULTS

Dov Basker

TABLE OF CONTENTS

I.   Introduction...126

II.   External Selection..126
        A.   Effect of Selection..126

III.   Internal Selection 127
        A.   Polygonal Tests 127
                1.   Three Products...129
                        a.   Secondary Identification...............129
                2.   Four Products...129
                3.   Five Products or More 129

IV.   General Correction for "Guessers".....................................129

V.   Chi-Square Evaluation Model...130

Appendix I: Correction of Preference Data Following a Triangular Test 130

Appendix II: General Data Correction Following A Difference Test 140

Appendix III: Chi-Square Model 141

References...143

No laboratory test can be severer than the market, which consists of millions of consumers. There is no appeal from their judgment.

— from an advertisement.

I. INTRODUCTION

One of the principal functions of industrial food analysis is to reduce the risk of encountering a negative market judgment. Within this framework, sensory analysis serves the same function with respect to sensory judgment.

Consider the problem of comparing two similar products in order to determine which is preferred by taste. Problems of this kind abound in industrial practice, when one ingredient is replaced by another, its concentration is changed, or a manufacturing process is altered. To examine the problem, a quantity of each of the two similar products will be required.

Scenario #1 calls for a few hundred potential consumers ("assessors"[1]) to taste samples of each of each of the two products, and to express their preference for one or the other. To simplify the interpretation, a "forced choice"[2] can be required of the assessors, disallowing tied preferences. An unknown proportion of the assessors are really unable to distinguish between the products, and that which these assessors record as being preferred is decided by a 50/50 chance. Among the more discriminating assessors, some will truly prefer one product (A) and some will prefer to the other (B): *de gustibus non disputandum* (there is no accounting for tastes). Statistical tables are available which facilitate the decision as to whether the total number of stated preferences for each of A and B is likely to have arisen by chance.[2]

Scenario #2 calls for a panel consisting of a dozen or so trained personnel to do the tasting. Because of their training and experience, a much smaller proportion of them is assumed to be really unable to differentiate between products A and B, and therefore fewer are actually "guessing" their preference. Scenario #3 calls for a single acknowledged expert to taste and decide; he presumably never simply guesses.

Scenario #3 is undoubtedly the simplest, quickest, and cheapest to perform, particularly as the expert is generally a staff member and may himself have suggested the comparison. However, the risk of idiosyncratic decision increases as the number of assessors decreases and attains a maximum with only one person.

II. EXTERNAL SELECTION

An alternative approach would be to screen the population of potential assessors, and to disqualify those less sensitive from taking part in the preference-decision process. As sensitivity to flavor nuances in one type of product is quite probably different from sensitivity in another type, the screening procedure needs to be product-specific. Potential assessors can be asked to try to distinguish between products in a triangular test:[3] three samples are presented simultaneously, labeled only with code numbers, and the assessors are told that two of the samples are in fact identical. Each assessor repeats this procedure several times with other similar products in the same quality range. All of the assessors, of course, need to compare the same products, and the frequencies with which they make correct identifications are compared.[4,5] The more sensitive assessors can thereafter be used as a panel for any later preference decisions. Admittedly, some may have been included "by accident" among this panel, and others may have been excluded equally accidentally, but the panel as a whole is more reliable than the unscreened population.

A. Effect of Selection

In theory, a finite difference exists between any two products, and can be detected only if a large enough population of assessors is used.[6] In practice, "a difference is only a difference if it makes a difference" — either to the consumers-at-large or to their opinion leaders, if any. As to the effect of selection, consider 20 experts who all clearly distinguish between products A and B, and 19 of whom prefer the former: no mathematics is required to conclude that their preference is fairly definitive. Yet if 980 undiscriminating assessors were added to the panel, the expected preferences combine to give 509:491, a result which could hardly be described as convincing.

Assessor selection cannot always be practiced, but where it is possible, the results thereby obtained generally well repay the effort.

III. INTERNAL SELECTION

It can be argued that a person's taste sensitivity changes with the passage of time and that a panel selected externally as above may not retain its advantages for long. It can also be argued that even for a given type of product, sensitivity over one quality range does not prove much regarding a different range.

This argument can be answered by presenting the available population of assessors with a triangular test of the actual samples whose preference-decision is required. The assessors are asked to try to identify the singlet and, having done so, to express their preference for either the pair or the singlet; the stated preferences of those assessors who do not make the correct identification can be ignored, thus eliminating two thirds of the "guessers". If the total number of assessors is not too small, a correction can be applied to the remaining preference data, to remove the effect of the remaining "guessers".[7] The method of calculating this correction is given in Appendix I.

The procedure described assists in two different ways. First, an estimate is obtained of the proportion of assessors able to distinguish between products A and B and of its significance; secondly, the precision of the remaining assessors' judgments is increased.

It was once reported that an experiment involving taste preference judgment following a triangular test appeared to show bias in favor of the sample pair and against the singlet.[8] Later work, however, did not substantiate the existence of such bias.[9] Nevertheless, some experimenters have tried to eliminate any such possibility by requiring the assessors to participate in two complementary triangular tests, with the pair and the singlet reversed, i.e. AAB and ABB ("back-to-back"[10]): the assessor's preference is considered only if the correct identification was made in both triangular tests and also if the same product (A or B) was preferred in both tests. Correct interpretation of such preference data must be based on the number of assessors who prefer, say, product A, and not on the number of assessments (which is twice as large); the logic of this statement is apparent upon consideration of extreme alternatives: greater reliability is clearly obtained from a single preference opinion of each of 100 assessors, than from 100 replicates by a single assessor.

A. Polygonal Tests

Many routine taste tests must perforce be carried out with relatively small panels, and perhaps with only one assessor (see Scenario #3 above). In such cases, increased statistical significance of difference identification can be obtained by replacing the traditional triangular test with a higher polygonal presentation.[11] For example, a 2 vs. 3 pentagonal test would have two samples of product A and three of product B. The odds that correct identifications will be made by chance are only one in ten, as compared to one in three for the triangular test. In a 3 vs. 4 heptagonal test, the odds that correct identifications will be made by chance decrease further to 1 in 35, while in a 4 vs. 5 nonagonal test the odds are only 1 in 126. Thus, the single expert who correctly identifies the samples in a nonagonal test demonstrates that he does so with appreciable significance ($p < 0.01$). Tables have been published for the interpretation of these polygonal tests, based on the cumulative binomial probability function (Tables 1A, 1B, and 1C in Basker[11]).

1. Three Products

When three products (A, B, and C) need to be compared, the use of the triangular or higher polygonal tests would increase the workload of a taste panel threefold, as each comparison of two products would require its own test. In practice, too, the conclusions may be mixed and confusing. A simultaneous solution can be found by the use of a 2 vs. 2 vs. 2 trihedral test,[11,12] where the odds of correctly identifying any one of these pairs by chance are one in five. The "trihedron" does not even require equal numbers of samples for A, B, and C, but in that case separate significance tables are required. Those assessors are identified who discriminate between each product and the remaining two.

a. Secondary Identification

If the panel as a whole clearly differentiates between one of three products (say, A) and the other two (B and C), a secondary identification level can be explored among those assessors who correctly identified the A samples. The primary discriminators of A then constitute a subpanel of higher sensitivity, but unless their number is sufficiently large, further differentiation between B and C is unlikely.

2. Four Products

A balanced tetrahedral test would require no less than $2 + 2 + 2 + 2 = 8$ samples to compare four products. Depending on their nature, they might strain the palate,[13,14] but even if fairly bland will strain an assessor's taste-memory. A 2 vs. 2 vs. 1 vs. 1 tetrahedral test[12] will decrease the effort required of the assessors while increasing that of the experimenter. Nevertheless, those assessors who discriminate each product among the others will be identified. If possible, secondary and even tertiary identifications levels may also be explored.

3. Five Products or More

No practical system of internal selection is presently available for the direct comparison of five or more products.

IV. GENERAL CORRECTION FOR "GUESSERS"

The method developed by Woodward and Schucany[7] for the correction of preference data following a triangular test (see Appendix I), may be extended to cover all forced-choice polygonal and polyhedral difference tests, and to cover quality scoring and nonparametric rating patterns as well. In principle, it is inferred that the results obtained from the assessors who failed to make a correct identification are reflected in those of the "lucky guessers" who made the identification correctly by chance.

In polygonal tests, preference is expressed for either product A or product B; the "lucky guessers" are considered equally likely to express preference for either product. In polyhedral tests, preferences are ranked in order (first preference = 1, second preference = 2, etc.), and totaling these rank values establishes their statistical significance level;[15] random rank equals $0.5 \times (1 + \text{number of products compared})$. Quality scores, on the other hand, are not distributed randomly evenly by "guessers", but are influenced by the actual qualities of the products; quality ratings are similarly not distributed randomly. Scores and nonparametric ratings calculated from incorrectly identified samples serve to be apportioned to the "lucky guessers". The methods of calculation are given in Appendix II.

V. CHI-SQUARE EVALUATION MODEL

Tables based on the cumulative binomial probability function have historically been used to evaluate the results of taste difference tests.[2] In Scenario #2 in the introduction, for example, if 75% of a 20-man panel prefer product A over product B, the $p = 0.05$ two-tailed statistical significance level has been attained for this function. Yet we should not be overly surprised if a replicate panel would result in the preference being equally divided ($\chi^2 = 2.2$, degrees of freedom $= 1$, $0.10 < p < 0.20$). The model for the initial decision considers the risk that the results might have been due simply to chance; the more exacting null hypothesis is that there is no preference between the products. This latter model may be extended to fit the general case of difference detection by triangular, polygonal, and polyhedral tests.

Minimum values (M) of the numbers of correct identifications required for statistical significance by the chi-square model have been computed for various common identification tests, and for the simple preference test of the previous paragraph. These values are given in Tables 1 to 5; when the number of assessors used (N) does not appear in these tables, M can be calculated from the values of d and e given in Table 6. The methods of calculation are detailed in Appendix III. The lower reaches of Tables 1 to 5 are curtailed in order that the theoretical contingency frequencies should be compatible with the chi-square approximation;[16] the binomial model is specifically indicated in such instances where experimental exigencies do not allow for the larger number of assessors required by the chi-square model.

Table 1 corresponds to the probability 1/2, and should be used for the simple (two-tailed) preference test. Table 2 corresponds to the probability 1/3 and should be used for the triangular test and for the singlets in a 1 vs. 1 vs. 2 vs. 2 tetrahedral test. Table 3 corresponds to the probability 1/5 and should be used for the 2 vs. 2 vs. 2 trihedral test. Table 4 corresponds to the probability 1/7.5 and should be used for the pairs in a 1 vs. 1 vs. 2 vs. 2 tetrahedral test. Table 5 corresponds to the probability 1/10 and should be used for the 2 vs. 3 pentagonal test.

The argument may be extended to detect possible nondifference of products,[17] and Tables 1 to 5 also give the maxima (M′) and minima (M″) for preferences (Table 1) and identifications. For values of N other than those listed, d', e', d'', and e'', are given in Table 6.

The argument may be extended yet further to cover identification tests where suspiciously few correct identifications are made. Here the possibility exists that the products compared indeed differ, but that the samples have been confused.[17] Values of the maximum numbers (M_o) of identifications are given in Tables 2 to 5, while for other values of N they can be calculated from d_o and e_o of Table 6.

"The proof of the pudding is in the eating."[18]

APPENDIX I:

CORRECTION OF PREFERENCE DATA FOLLOWING A TRIANGULAR TEST

N_t = total number of assessors
N_c = number of assessors who correctly identify the singlet
P_a = number of assessors, of N_c, who prefer product A
P_b = number of assessors, of N_c, who prefer product B

then the corrected numbers of preferences are

$$P'_a = P_a - 0.25\ N_t + 0.25\ N_c$$

and $\quad P'_b = P_b - 0.25\ N_t + 0.25\ N_c$

of $\quad N'_c = P'_a + P'_b$

Table 1

STATISTICAL SIGNIFICANCE OF TASTE PANEL TEST RESULTS, BASED ON THE CHI-SQUARE MODEL[a]

Random Probability $= 1/2$

	Probability								
	M″ and M′						**M**		
N	0.95		0.99		0.999		0.05	0.01	0.001
9	—		—		—		—	—	—
10	5 −	5	5 −	5	5 −	5	—	—	—
11	5 −	6	5 −	6	5 −	6	—	—	—
12	6 −	6	6 −	6	6 −	6	—	—	—
13	6 −	7	6 −	7	6 −	7	—	—	—
14	6 −	8	7 −	7	7 −	7	—	—	—
15	7 −	8	7 −	8	7 −	8	—	—	—
16	7 −	9	8 −	8	8 −	8	—	—	—
17	8 −	9	8 −	9	8 −	9	15	—	—
18	8 −	10	9 −	9	9 −	9	16	—	—
19	9 −	10	9 −	10	9 −	10	17	18	—
20	9 −	11	9 −	11	10 −	10	17	19	—
21	10 −	11	10 −	11	10 −	11	18	20	—
22	10 −	12	10 −	12	10 −	12	19	20	—
23	11 −	12	11 −	12	11 −	12	19	21	23
24	11 −	13	11 −	13	11 −	13	20	22	24
25	12 −	13	12 −	13	12 −	13	21	22	25
26	12 −	14	12 −	14	12 −	14	21	23	25
27	13 −	14	13 −	14	13 −	14	22	24	26
28	13 −	15	13 −	15	13 −	15	22	25	27
29	14 −	15	14 −	15	14 −	15	23	25	28
30	14 −	16	14 −	16	14 −	16	24	26	28
31	15 −	16	15 −	16	15 −	16	24	27	29
32	15 −	17	15 −	17	15 −	17	25	27	30
33	16 −	17	16 −	17	16 −	17	26	28	30
34	16 −	18	16 −	18	16 −	18	26	29	31
35	17 −	18	17 −	18	17 −	18	27	29	32
36	17 −	19	17 −	19	17 −	19	28	30	32
37	18 −	19	18 −	19	18 −	19	28	31	33
38	18 −	20	18 −	20	18 −	20	29	31	34
39	19 −	20	19 −	20	19 −	20	29	32	35
40	19 −	21	19 −	21	19 −	21	30	32	35
41	20 −	21	20 −	21	20 −	21	31	33	36
42	20 −	22	20 −	22	20 −	22	31	34	37
43	21 −	22	21 −	22	21 −	22	32	34	37
44	21 −	23	21 −	23	21 −	23	32	35	38
45	22 −	23	22 −	23	22 −	23	33	36	39
46	22 −	24	22 −	24	22 −	24	34	36	39
47	23 −	24	23 −	24	23 −	24	34	37	40
48	23 −	25	23 −	25	23 −	25	35	38	41
49	24 −	25	24 −	25	24 −	25	35	38	41
50	24 −	26	24 −	26	24 −	26	36	39	42
51	25 −	26	25 −	26	25 −	26	37	39	43
52	25 −	27	25 −	27	25 −	27	37	40	43
53	26 −	27	26 −	27	26 −	27	38	41	44
54	26 −	28	26 −	28	26 −	28	38	41	45
55	27 −	28	27 −	28	27 −	28	39	42	45
56	27 −	29	27 −	29	27 −	29	40	43	46
57	28 −	29	28 −	29	28 −	29	40	43	47
58	28 −	30	28 −	30	28 −	30	41	44	47

Table 1 (continued)
STATISTICAL SIGNIFICANCE OF TASTE PANEL TEST
RESULTS, BASED ON THE CHI-SQUARE MODEL[a]

Random Probability = $^1/_2$

Probability

N	M″ and M′						M		
	0.95		0.99		0.999		0.05	0.01	0.001
59	29 −	30	29 −	30	29 −	30	41	44	48
60	29 −	31	29 −	31	29 −	31	42	45	49
61	30 −	31	30 −	31	30 −	31	43	46	49
62	30 −	32	30 −	32	30 −	32	43	46	50
63	31 −	32	31 −	32	31 −	32	44	47	51
64	31 −	33	31 −	33	31 −	33	44	48	51
65	32 −	33	32 −	33	32 −	33	45	48	52
66	32 −	34	32 −	34	32 −	34	46	49	52
67	33 −	34	33 −	34	33 −	34	46	49	53
68	33 −	35	33 −	35	33 −	35	47	50	54
69	34 −	35	34 −	35	34 −	35	47	51	54
70	34 −	36	34 −	36	34 −	36	48	51	55
71	35 −	36	35 −	36	35 −	36	49	52	56
72	35 −	37	35 −	37	35 −	37	49	52	56
73	36 −	37	36 −	37	36 −	37	50	53	57
74	36 −	38	36 −	38	36 −	38	50	54	58
75	37 −	38	37 −	38	37 −	38	51	54	58
76	37 −	39	37 −	39	37 −	39	51	55	59
77	38 −	39	38 −	39	38 −	39	52	56	60
78	38 −	40	38 −	40	38 −	40	53	56	60
79	39 −	40	39 −	40	39 −	40	53	57	61
80	39 −	41	39 −	41	39 −	41	54	57	61
81	40 −	41	40 −	41	40 −	41	54	58	62
82	40 −	42	40 −	42	40 −	42	55	59	63
83	41 −	42	41 −	42	41 −	42	55	59	63
84	41 −	43	41 −	43	41 −	43	56	60	64
85	42 −	43	42 −	43	42 −	43	57	60	65
86	42 −	44	42 −	44	42 −	44	57	61	65
87	43 −	44	43 −	44	43 −	44	58	62	66
88	43 −	45	43 −	45	43 −	45	58	62	66
89	44 −	45	44 −	45	44 −	45	59	63	67
90	44 −	46	44 −	46	44 −	46	60	63	68
91	45 −	46	45 −	46	45 −	46	60	64	68
92	45 −	47	45 −	47	45 −	47	61	65	69
93	46 −	47	46 −	47	46 −	47	61	65	70
94	46 −	48	46 −	48	46 −	48	62	66	70
95	47 −	48	47 −	48	47 −	48	62	66	71
96	47 −	49	47 −	49	47 −	49	63	67	71
97	47 −	50	48 −	49	48 −	49	64	68	72
98	48 −	50	48 −	50	48 −	50	64	68	73
99	48 −	51	49 −	50	49 −	50	65	69	73
100	49 −	51	49 −	51	49 −	51	65	69	74
200	99 −	101	99 −	101	99 −	101	121	127	134
300	149 −	151	149 −	151	149 −	151	175	183	191
400	199 −	201	199 −	201	199 −	201	229	238	248
500	249 −	251	249 −	251	249 −	251	282	292	303
750	373 −	377	374 −	376	374 −	376	414	426	440
1000	498 −	502	499 −	501	499 −	501	545	559	575
1500	748 −	752	749 −	751	749 −	751	805	822	841
2000	998 −	1002	999 −	1001	999 −	1001	1063	1083	1105

[a] See text for nomenclature.

Table 2
STATISTICAL SIGNIFICANCE OF TASTE PANEL TEST RESULTS BASED
ON THE CHI-SQUARE MODEL[a]

Random Probability = $^1/_3$

Probability

N	M_o			M'' and M'			M		
	0.001	0.01	0.05	0.95	0.99	0.999	0.05	0.01	0.001
11	—	—	—	——	——	——	—	—	—
12	—	—	—	——	——	——	10	—	—
13	—	—	—	——	——	——	11	—	—
14	—	—	—	——	——	——	11	13	—
15	—	—	—	5 – 5	5 – 5	5 – 5	12	13	15
16	—	—	—	5 – 6	5 – 6	5 – 6	12	14	16
17	—	—	—	5 – 6	5 – 6	5 – 6	13	15	17
18	—	—	—	5 – 7	6 – 6	6 – 6	13	15	17
19	—	—	—	6 – 7	6 – 7	6 – 7	14	16	18
20	—	—	—	6 – 7	6 – 7	6 – 7	14	16	18
21	—	—	—	6 – 8	6 – 8	7 – 7	15	17	19
22	—	—	—	7 – 8	7 – 8	7 – 8	15	17	20
23	—	—	—	7 – 8	7 – 8	7 – 8	16	18	20
24	—	—	—	7 – 9	7 – 9	7 – 9	16	18	21
25	—	—	—	8 – 9	8 – 9	8 – 9	17	19	21
26	—	—	—	8 – 9	8 – 9	8 – 9	17	19	22
27	—	—	2	8 – 10	8 – 10	8 – 10	18	20	22
28	—	—	2	9 – 10	9 – 10	9 – 10	18	20	23
29	—	—	2	9 – 10	9 – 10	9 – 10	19	21	24
30	—	0	2	9 – 11	9 – 11	9 – 11	19	21	24
31	0	1	2	10 – 11	10 – 11	10 – 11	19	22	25
32	0	1	3	10 – 11	10 – 11	10 – 11	20	22	25
33	0	1	3	10 – 12	10 – 12	10 – 12	20	23	26
34	0	1	3	11 – 12	11 – 12	11 – 12	21	23	26
35	0	1	3	11 – 12	11 – 12	11 – 12	21	24	27
36	0	2	4	11 – 13	11 – 13	11 – 13	22	24	27
37	0	2	4	12 – 13	12 – 13	12 – 13	22	25	28
38	0	2	4	12 – 13	12 – 13	12 – 13	23	25	28
39	0	2	4	12 – 14	12 – 14	12 – 14	23	26	29
40	0	2	4	13 – 14	13 – 14	13 – 14	24	26	29
41	0	2	5	13 – 14	13 – 14	13 – 14	24	27	30
42	0	3	5	13 – 15	13 – 15	13 – 15	24	27	30
43	1	3	5	14 – 15	14 – 15	14 – 15	25	28	31
44	1	3	5	14 – 15	14 – 15	14 – 15	25	28	32
45	1	3	6	14 – 16	14 – 16	14 – 16	26	29	32
46	1	4	6	15 – 16	15 – 16	15 – 16	26	29	33
47	1	4	6	15 – 16	15 – 16	15 – 16	27	30	33
48	1	4	6	15 – 17	15 – 17	15 – 17	27	30	34
49	2	4	7	16 – 17	16 – 17	16 – 17	27	31	34
50	2	4	7	16 – 17	16 – 17	16 – 17	28	31	35
51	2	5	7	16 – 18	16 – 18	16 – 18	28	31	35
52	2	5	7	17 – 18	17 – 18	17 – 18	29	32	36
53	2	5	7	17 – 18	17 – 18	17 – 18	29	32	36
54	2	5	8	17 – 19	17 – 19	17 – 19	30	33	37
55	3	5	8	18 – 19	18 – 19	18 – 19	30	33	37
56	3	6	8	18 – 19	18 – 19	18 – 19	30	34	37
57	3	6	8	18 – 20	18 – 20	18 – 20	31	34	38
58	3	6	9	18 – 20	19 – 20	19 – 20	31	35	38
59	3	6	9	19 – 21	19 – 20	19 – 20	32	35	39
60	4	7	9	19 – 21	19 – 21	19 – 21	32	36	39

Table 2 (continued)
STATISTICAL SIGNIFICANCE OF TASTE PANEL TEST RESULTS BASED ON THE CHI-SQUARE MODEL[a]

Random Probability $= 1/_3$

	Probability								
	M_o			M″ and M′			M		
N	0.001	0.01	0.05	0.95	0.99	0.999	0.05	0.01	0.001
61	4	7	9	19 − 21	20 − 21	20 − 21	33	36	40
62	4	7	10	20 − 22	20 − 21	20 − 21	33	36	40
63	4	7	10	20 − 22	20 − 22	20 − 22	33	37	41
64	4	7	10	20 − 22	21 − 22	21 − 22	34	37	41
65	5	8	10	21 − 23	21 − 22	21 − 22	34	38	42
66	5	8	11	21 − 23	21 − 23	21 − 23	35	38	42
67	5	8	11	21 − 23	22 − 23	22 − 23	35	39	43
68	5	8	11	22 − 24	22 − 23	22 − 23	36	39	43
69	5	9	11	22 − 24	22 − 24	22 − 24	36	40	44
70	6	9	12	22 − 24	23 − 24	23 − 24	36	40	44
71	6	9	12	23 − 25	23 − 24	23 − 24	37	40	45
72	6	9	12	23 − 25	23 − 25	23 − 25	37	41	45
73	6	10	12	23 − 25	24 − 25	24 − 25	38	41	46
74	6	10	13	24 − 26	24 − 25	24 − 25	38	42	46
75	7	10	13	24 − 26	24 − 26	24 − 26	38	42	47
76	7	10	13	24 − 26	25 − 26	25 − 26	39	43	47
77	7	10	14	25 − 27	25 − 26	25 − 26	39	43	47
78	7	11	14	25 − 27	25 − 27	25 − 27	40	44	48
79	8	11	14	25 − 27	26 − 27	26 − 27	40	44	48
80	8	11	14	26 − 28	26 − 27	26 − 27	41	44	49
81	8	11	15	26 − 28	26 − 28	26 − 28	41	45	49
82	8	12	15	26 − 28	27 − 28	27 − 28	41	45	50
83	8	12	15	27 − 29	27 − 28	27 − 28	42	46	50
84	9	12	15	27 − 29	27 − 29	27 − 29	42	46	51
85	9	12	16	27 − 29	28 − 29	28 − 29	43	47	51
86	9	13	16	28 − 30	28 − 29	28 − 29	43	47	52
87	9	13	16	28 − 30	28 − 30	28 − 30	43	47	52
88	9	13	16	28 − 30	29 − 30	29 − 30	44	48	53
89	10	13	17	29 − 31	29 − 30	29 − 30	44	48	53
90	10	14	17	29 − 31	29 − 31	29 − 31	45	49	53
91	10	14	17	29 − 31	30 − 31	30 − 31	45	49	54
92	10	14	17	30 − 32	30 − 31	30 − 31	45	50	54
93	11	14	18	30 − 32	30 − 32	30 − 32	46	50	55
94	11	15	18	30 − 32	31 − 32	31 − 32	46	50	55
95	11	15	18	31 − 33	31 − 32	31 − 32	47	51	56
96	11	15	19	31 − 33	31 − 33	31 − 33	47	51	56
97	11	15	19	31 − 33	32 − 33	32 − 33	47	52	57
98	12	16	19	32 − 34	32 − 33	32 − 33	48	52	57
99	12	16	19	32 − 34	32 − 34	32 − 34	48	53	58
100	12	16	20	32 − 34	33 − 34	33 − 34	49	53	58
200	36	42	47	66 − 68	66 − 67	66 − 67	87	93	101
300	63	70	77	99 − 101	99 − 101	99 − 101	125	132	141
400	90	99	106	132 − 135	133 − 134	133 − 134	162	170	180
500	118	128	137	165 − 168	166 − 167	166 − 167	198	208	219
750	190	203	213	248 − 252	249 − 251	249 − 251	288	300	313
1000	264	279	291	332 − 335	333 − 334	333 − 334	377	390	406
1500	415	433	449	498 − 502	499 − 501	499 − 501	553	569	588
2000	569	590	607	664 − 669	666 − 668	666 − 667	727	746	768

[a] See text for nomenclature.

Table 3
STATISTICAL SIGNIFICANCE OF TASTE PANEL TEST RESULTS, BASED ON THE CHI-SQUARE MODEL[a]

Random Probability $= \frac{1}{5}$

				Probability						
		M_o			M'' and M'				M	
N	0.001	0.01	0.05	0.95	0.99	0.999		0.05	0.01	0.001
10	—	—	—	——	——	——		—	—	—
11	—	—	—	——	——	——		8	9	—
12	—	—	—	——	——	——		8	10	—
13	—	—	—	——	——	——		9	10	12
14	—	—	—	——	——	——		9	11	13
15	—	—	—	——	——	——		9	11	13
16	—	—	—	——	——	——		10	12	14
17	—	—	—	——	——	——		10	12	14
18	—	—	—	——	——	——		11	12	15
19	—	—	—	——	——	——		11	13	15
20	—	—	—	——	——	——		11	13	16
21	—	—	—	——	——	——		12	14	16
22	—	—	—	——	——	——		12	14	17
23	—	—	—	——	——	——		12	14	17
24	—	—	—	——	——	——		13	15	17
25	—	—	—	——	——	——		13	15	18
26	—	—	—	5 – 6	5 – 6	5 – 6		13	16	18
27	—	—	—	5 – 6	5 – 6	5 – 6		14	16	19
28	—	—	—	5 – 6	5 – 6	5 – 6		14	16	19
29	—	—	—	5 – 6	5 – 6	5 – 6		14	17	20
30	—	—	—	5 – 7	5 – 7	5 – 7		15	17	20
31	—	—	—	6 – 7	6 – 7	6 – 7		15	17	20
32	—	—	—	6 – 7	6 – 7	6 – 7		15	18	21
33	—	—	—	6 – 7	6 – 7	6 – 7		15	18	21
34	—	—	—	6 – 7	6 – 7	6 – 7		16	18	21
35	—	—	—	6 – 8	6 – 8	6 – 8		16	19	22
36	—	—	—	7 – 8	7 – 8	7 – 8		16	19	22
37	—	—	—	7 – 8	7 – 8	7 – 8		17	19	23
38	—	—	—	7 – 8	7 – 8	7 – 8		17	20	23
39	—	—	—	7 – 9	7 – 8	7 – 8		17	20	23
40	—	—	—	7 – 9	7 – 9	7 – 9		18	20	24
41	—	—	—	7 – 9	8 – 9	8 – 9		18	21	24
42	—	—	—	8 – 9	8 – 9	8 – 9		18	21	24
43	—	—	—	8 – 9	8 – 9	8 – 9		18	21	25
44	—	—	—	8 – 10	8 – 9	8 – 9		19	22	25
45	—	—	1	8 – 10	8 – 10	8 – 10		19	22	26
46	—	—	1	8 – 10	9 – 10	9 – 10		19	22	26
47	—	—	2	9 – 10	9 – 10	9 – 10		20	23	26
48	—	—	2	9 – 10	9 – 10	9 – 10		20	23	27
49	—	—	2	9 – 11	9 – 10	9 – 10		20	23	27
50	—	0	2	9 – 11	9 – 11	9 – 11		21	24	27
51	—	0	2	9 – 11	10 – 11	10 – 11		21	24	28
52	0	0	2	10 – 11	10 – 11	10 – 11		21	24	28
53	0	0	2	10 – 11	10 – 11	10 – 11		21	25	28
54	0	1	2	10 – 12	10 – 11	10 – 11		22	25	29
55	0	1	3	10 – 12	10 – 12	10 – 12		22	25	29
56	0	1	3	10 – 12	11 – 12	11 – 12		22	26	29
57	0	1	3	11 – 12	11 – 12	11 – 12		23	26	30
58	0	1	3	11 – 12	11 – 12	11 – 12		23	26	30

Table 3 (continued)
STATISTICAL SIGNIFICANCE OF TASTE PANEL TEST RESULTS, BASED ON THE CHI-SQUARE MODEL[a]

Random Probability $= {}^1/_5$

Probability

N	M_o 0.001	0.01	0.05	M″ and M′ 0.95	0.99	0.999	M 0.05	0.01	0.001
59	0	1	3	11 − 13	11 − 12	11 − 12	23	26	30
60	0	1	3	11 − 13	11 − 13	11 − 13	23	27	31
61	0	1	3	11 − 13	12 − 13	12 − 13	24	27	31
62	0	1	3	12 − 13	12 − 13	12 − 13	24	27	31
63	0	2	4	12 − 13	12 − 13	12 − 13	24	28	32
64	0	2	4	12 − 14	12 − 13	12 − 13	24	28	32
65	0	2	4	12 − 14	12 − 14	12 − 14	25	28	32
66	0	2	4	12 − 14	13 − 14	13 − 14	25	29	33
67	0	2	4	13 − 14	13 − 14	13 − 14	25	29	33
68	0	2	4	13 − 14	13 − 14	13 − 14	26	29	33
69	0	2	4	13 − 15	13 − 14	13 − 14	26	29	34
70	0	2	4	13 − 15	13 − 15	13 − 15	26	30	34
71	0	2	5	13 − 15	14 − 15	14 − 15	26	30	34
72	0	3	5	14 − 15	14 − 15	14 − 15	27	30	35
73	0	3	5	14 − 15	14 − 15	14 − 15	27	31	35
74	1	3	5	14 − 16	14 − 15	14 − 15	27	31	35
75	1	3	5	14 − 16	14 − 16	14 − 16	27	31	36
76	1	3	5	14 − 16	15 − 16	15 − 16	28	31	36
77	1	3	5	15 − 16	15 − 16	15 − 16	28	32	36
78	1	3	6	15 − 16	15 − 16	15 − 16	28	32	37
79	1	3	6	15 − 17	15 − 16	15 − 16	29	32	37
80	1	4	6	15 − 17	15 − 17	15 − 17	29	33	37
81	1	4	6	15 − 17	16 − 17	16 − 17	29	33	38
82	1	4	6	16 − 17	16 − 17	16 − 17	29	33	38
83	1	4	6	16 − 17	16 − 17	16 − 17	30	34	38
84	2	4	6	16 − 18	16 − 17	16 − 17	30	34	38
85	2	4	7	16 − 18	16 − 18	16 − 18	30	34	39
86	2	4	7	16 − 18	17 − 18	17 − 18	30	34	39
87	2	4	7	17 − 18	17 − 18	17 − 18	31	35	39
88	2	5	7	17 − 18	17 − 18	17 − 18	31	35	40
89	2	5	7	17 − 19	17 − 18	17 − 18	31	35	40
90	2	5	7	17 − 19	17 − 19	17 − 19	31	35	40
91	2	5	7	17 − 19	18 − 19	18 − 19	32	36	41
92	2	5	8	18 − 19	18 − 19	18 − 19	32	36	41
93	3	5	8	18 − 20	18 − 19	18 − 19	32	36	41
94	3	5	8	18 − 20	18 − 19	18 − 19	32	37	42
95	3	5	8	18 − 20	18 − 20	18 − 20	33	37	42
96	3	6	8	18 − 20	19 − 20	19 − 20	33	37	42
97	3	6	8	19 − 20	19 − 20	19 − 20	33	37	42
98	3	6	8	19 − 21	19 − 20	19 − 20	34	38	43
99	3	6	9	19 − 21	19 − 20	19 − 20	34	38	43
100	3	6	9	19 − 21	19 − 21	19 − 21	34	38	43
200	16	20	24	39 − 41	39 − 41	39 − 41	58	64	71
300	30	35	41	59 − 61	59 − 61	59 − 61	82	89	97
400	45	51	58	79 − 81	79 − 81	79 − 81	105	113	122
500	60	68	75	99 − 101	99 − 101	99 − 101	127	136	146
750	101	111	119	149 − 151	149 − 151	149 − 151	183	193	206
1000	143	154	165	198 − 202	199 − 201	199 − 201	238	250	264
1500	230	244	257	298 − 302	299 − 301	299 − 301	346	360	377
2000	319	335	350	398 − 402	399 − 401	399 − 401	452	469	488

[a] See text for nomenclature.

Table 4
STATISTICAL SIGNIFICANCE OF TASTE PANEL TEST RESULTS, BASED ON THE CHI-SQUARE MODEL[a]

Random Probability = $^1/_{7.5}$

Probability

N	M$_o$			M'' and M'			M		
	0.001	0.01	0.05	0.95	0.99	0.999	0.05	0.01	0.001
10	—	—	—	——	——	——	—	—	
11	—	—	—	——	——	——	—	9	—
12	—	—	—	——	——	——	—	9	11
13	—	—	—	——	——	——	—	9	11
14	—	—	—	——	——	——	—	10	12
15	—	—	—	——	——	——	8	10	12
16	—	—	—	——	——	——	8	10	13
17	—	—	—	——	——	——	9	11	13
18	—	—	—	——	——	——	9	11	13
19	—	—	—	——	——	——	9	11	14
20	—	—	—	——	——	——	9	12	14
21	—	—	—	——	——	——	10	12	14
22	—	—	—	——	——	——	10	12	15
23	—	—	—	——	——	——	10	12	15
24	—	—	—	——	——	——	11	13	15
25	—	—	—	——	——	——	11	13	16
26	—	—	—	——	——	——	11	13	16
27	—	—	—	——	——	——	11	14	16
28	—	—	—	——	——	——	12	14	17
29	—	—	—	——	——	——	12	14	17
30	—	—	—	——	——	——	12	14	17
31	—	—	—	——	——	——	12	15	18
32	—	—	—	——	——	——	12	15	18
33	—	—	—	——	——	——	13	15	18
34	—	—	—	——	——	——	13	15	19
35	—	—	—	——	——	——	13	16	19
36	—	—	—	——	——	——	13	16	19
37	—	—	—	——	——	——	14	16	20
38	—	—	—	——	5– 6	5– 6	14	17	20
39	—	—	—	5– 6	5– 6	5– 6	14	17	20
40	—	—	—	5– 6	5– 6	5– 6	14	17	20
41	—	—	—	5– 6	5– 6	5– 6	14	17	21
42	—	—	—	5– 6	5– 6	5– 6	15	18	21
43	—	—	—	5– 6	5– 6	5– 6	15	18	21
44	—	—	—	5– 7	5– 6	5– 6	15	18	21
45	—	—	—	5– 7	5– 7	5– 7	15	18	22
46	—	—	—	5– 7	6– 7	6– 7	16	18	22
47	—	—	—	6– 7	6– 7	6– 7	16	19	22
48	—	—	—	6– 7	6– 7	6– 7	16	19	23
49	—	—	—	6– 7	6– 7	6– 7	16	19	23
50	—	—	—	6– 7	6– 7	6– 7	16	19	23
51	—	—	—	6– 8	6– 7	6– 7	17	20	23
52	—	—	—	6– 8	6– 7	6– 7	17	20	24
53	—	—	—	6– 8	7– 8	7– 8	17	20	24
54	—	—	—	6– 8	7– 8	7– 8	17	20	24
55	—	—	—	7– 8	7– 8	7– 8	17	21	24
56	—	—	—	7– 8	7– 8	7– 8	18	21	25
57	—	—	—	7– 8	7– 8	7– 8	18	21	25
58	—	—	—	7– 8	7– 8	7– 8	18	21	25
59	—	—	—	7– 9	7– 8	7– 8	18	22	25

Table 4 (continued)
STATISTICAL SIGNIFICANCE OF TASTE PANEL TEST RESULTS, BASED ON THE CHI-SQUARE MODEL[a]

Random Probability $= 1/_{7.5}$

Probability

N	M_o 0.001	0.01	0.05	M" and M' 0.95	0.99	0.999	M 0.05	0.01	0.001
60	—	—	—	7 – 9	7 – 9	7 – 9	18	22	26
61	—	—	—	7 – 9	8 – 9	8 – 9	19	22	26
62	—	—	—	8 – 9	8 – 9	8 – 9	19	22	26
63	—	—	—	8 – 9	8 – 9	8 – 9	19	22	26
64	—	—	—	8 – 9	8 – 9	8 – 9	19	23	27
65	—	—	—	8 – 9	8 – 9	8 – 9	19	23	27
66	—	—	—	8 – 10	8 – 9	8 – 9	20	23	27
67	—	—	—	8 – 10	8 – 9	8 – 9	20	23	27
68	—	—	1	8 – 10	9 – 10	9 – 10	20	23	28
69	—	—	1	8 – 10	9 – 10	9 – 10	20	24	28
70	—	—	1	9 – 10	9 – 10	9 – 10	20	24	28
71	—	—	2	9 – 10	9 – 10	9 – 10	21	24	28
72	—	—	2	9 – 10	9 – 10	9 – 10	21	24	29
73	—	—	2	9 – 11	9 – 10	9 – 10	21	25	29
74	—	—	2	9 – 11	9 – 10	9 – 10	21	25	29
75	—	0	2	9 – 11	9 – 11	9 – 11	21	25	29
76	0	0	2	9 – 11	10 – 11	10 – 11	22	25	30
77	0	0	2	9 – 11	10 – 11	10 – 11	22	25	30
78	0	0	2	10 – 11	10 – 11	10 – 11	22	26	30
79	0	0	2	10 – 11	10 – 11	10 – 11	22	26	30
80	0	0	2	10 – 11	10 – 11	10 – 11	22	26	31
81	0	1	2	10 – 12	10 – 11	10 – 11	23	26	31
82	0	1	2	10 – 12	10 – 11	10 – 11	23	26	31
83	0	1	2	10 – 12	11 – 12	11 – 12	23	27	31
84	0	1	3	10 – 12	11 – 12	11 – 12	23	27	31
85	0	1	3	11 – 12	11 – 12	11 – 12	23	27	32
86	0	1	3	11 – 12	11 – 12	11 – 12	24	27	32
87	0	1	3	11 – 12	11 – 12	11 – 12	24	28	32
88	0	1	3	11 – 13	11 – 12	11 – 12	24	28	32
89	0	1	3	11 – 13	11 – 12	11 – 12	24	28	33
90	0	1	3	11 – 13	11 – 13	11 – 13	24	28	33
91	0	1	3	11 – 13	12 – 13	12 – 13	24	28	33
92	0	1	3	11 – 13	12 – 13	12 – 13	25	29	33
93	0	1	3	12 – 13	12 – 13	12 – 13	25	29	33
94	0	1	3	12 – 13	12 – 13	12 – 13	25	29	34
95	0	2	3	12 – 14	12 – 13	12 – 13	25	29	34
96	0	2	4	12 – 14	12 – 13	12 – 13	25	29	34
97	0	2	4	12 – 14	12 – 14	12 – 13	26	30	34
98	0	2	4	12 – 14	12 – 14	13 – 14	26	30	35
99	0	2	4	12 – 14	13 – 14	13 – 14	26	30	35
100	0	2	4	12 – 14	13 – 14	13 – 14	26	30	35
200	7	10	13	26 – 28	26 – 27	26 – 27	43	48	55
300	15	20	24	39 – 41	39 – 41	39 – 41	59	65	73
400	24	30	34	52 – 54	53 – 54	53 – 54	75	82	91
500	34	40	46	65 – 68	66 – 67	66 – 67	91	98	108
750	59	67	74	99 – 101	99 – 101	99 – 101	129	138	149
1000	86	95	103	132 – 135	133 – 134	133 – 134	166	176	189
1500	141	153	163	198 – 202	199 – 201	199 – 201	239	252	267
2000	198	212	224	265 – 269	266 – 267	266 – 267	312	326	343

[a] See text for nomenclature.

Table 5
STATISTICAL SIGNIFICANCE OF TASTE PANEL TEST RESULTS, BASED ON THE CHI-SQUARE MODEL[a]

Probability $= \frac{1}{10}$

Probabiity

N	M$_o$			M'' and M'			M		
	0.001	0.01	0.05	0.95	0.99	0.999	0.05	0.01	0.001
10	—	—	—	—	—	—	—	—	—
11	—	—	—	—	—	—	—	—	10
12	—	—	—	—	—	—	—	—	10
13	—	—	—	—	—	—	—	9	11
14	—	—	—	—	—	—	—	9	11
15	—	—	—	—	—	—	—	9	12
16	—	—	—	—	—	—	—	10	12
17	—	—	—	—	—	—	—	10	12
18	—	—	—	—	—	—	—	10	13
19	—	—	—	—	—	—	—	10	13
20	—	—	—	—	—	—	9	11	13
21	—	—	—	—	—	—	9	11	13
22	—	—	—	—	—	—	9	11	14
23	—	—	—	—	—	—	9	11	14
24	—	—	—	—	—	—	9	12	14
25	—	—	—	—	—	—	10	12	15
26	—	—	—	—	—	—	10	12	15
27	—	—	—	—	—	—	10	12	15
28	—	—	—	—	—	—	10	13	15
29	—	—	—	—	—	—	10	13	16
30	—	—	—	—	—	—	11	13	16
31	—	—	—	—	—	—	11	13	16
32	—	—	—	—	—	—	11	13	17
33	—	—	—	—	—	—	11	14	17
34	—	—	—	—	—	—	11	14	17
35	—	—	—	—	—	—	12	14	17
36	—	—	—	—	—	—	12	14	18
37	—	—	—	—	—	—	12	15	18
38	—	—	—	—	—	—	12	15	18
39	—	—	—	—	—	—	12	15	18
40	—	—	—	—	—	—	12	15	18
41	—	—	—	—	—	—	13	15	19
42	—	—	—	—	—	—	13	16	19
43	—	—	—	—	—	—	13	16	19
44	—	—	—	—	—	—	13	16	19
45	—	—	—	—	—	—	13	16	20
46	—	—	—	—	—	—	14	16	20
47	—	—	—	—	—	—	14	17	20
48	—	—	—	—	—	—	14	17	20
49	—	—	—	—	—	—	14	17	21
50	—	—	—	—	—	—	14	17	21
51	—	—	—	5– 6	5– 6	5– 6	14	17	21
52	—	—	—	5– 6	5– 6	5– 6	15	18	21
53	—	—	—	5– 6	5– 6	5– 6	15	18	21
54	—	—	—	5– 6	5– 6	5– 6	15	18	22
55	—	—	—	5– 6	5– 6	5– 6	15	18	22
56	—	—	—	5– 6	5– 6	5– 6	15	18	22
57	—	—	—	5– 6	5– 6	5– 6	15	18	22
58	—	—	—	5– 7	5– 6	5– 6	16	19	22
59	—	—	—	5– 7	5– 6	5– 6	16	19	23

Table 5 (continued)
STATISTICAL SIGNIFICANCE OF TASTE PANEL TEST RESULTS, BASED ON THE CHI-SQUARE MODEL[a]

Probability $= \frac{1}{10}$

				Probabiity							
		M_o			M'' and M'					M	
N	0.001	0.01	0.05	0.95		0.99		0.999	0.05	0.01	0.001
60	—	—	—	5 − 7	5 − 7		5 − 7		16	19	23
61	—	—	—	5 − 7	6 − 7		6 − 7		16	19	23
62	—	—	—	5 − 7	6 − 7		6 − 7		16	19	23
63	—	—	—	6 − 7	6 − 7		6 − 7		16	20	24
64	—	—	—	6 − 7	6 − 7		6 − 7		16	20	24
65	—	—	—	6 − 7	6 − 7		6 − 7		17	20	24
66	—	—	—	6 − 7	6 − 7		6 − 7		17	20	24
67	—	—	—	6 − 7	6 − 7		6 − 7		17	20	24
68	—	—	—	6 − 8	6 − 7		6 − 7		17	20	25
69	—	—	—	6 − 8	6 − 7		6 − 7		17	21	25
70	—	—	—	6 − 8	6 − 8		6 − 8		17	21	25
71	—	—	—	6 − 8	7 − 8		7 − 8		18	21	25
72	—	—	—	6 − 8	7 − 8		7 − 8		18	21	25
73	—	—	—	7 − 8	7 − 8		7 − 8		18	21	26
74	—	—	—	7 − 8	7 − 8		7 − 8		18	21	26
75	—	—	—	7 − 8	7 − 8		7 − 8		18	22	26
76	—	—	—	7 − 8	7 − 8		7 − 8		18	22	26
77	—	—	—	7 − 8	7 − 8		7 − 8		19	22	26
78	—	—	—	7 − 9	7 − 8		7 − 8		19	22	26
79	—	—	—	7 − 9	7 − 8		7 − 8		19	22	27
80	—	—	—	7 − 9	7 − 9		7 − 9		19	23	27
81	—	—	—	7 − 9	8 − 9		8 − 9		19	23	27
82	—	—	—	7 − 9	8 − 9		8 − 9		19	23	27
83	—	—	—	8 − 9	8 − 9		8 − 9		19	23	27
84	—	—	—	8 − 9	8 − 9		8 − 9		20	23	28
85	—	—	—	8 − 9	8 − 9		8 − 9		20	23	28
86	—	—	—	8 − 9	8 − 9		8 − 9		20	24	28
87	—	—	—	8 − 9	8 − 9		8 − 9		20	24	28
88	—	—	—	8 − 10	8 − 9		8 − 9		20	24	28
89	—	—	—	8 − 10	8 − 9		8 − 9		20	24	29
90	—	—	1	8 − 10	8 − 10		8 − 10		20	24	29
91	—	—	1	8 − 10	9 − 10		9 − 10		21	24	29
92	—	—	1	8 − 10	9 − 10		9 − 10		21	25	29
93	—	—	1	9 − 10	9 − 10		9 − 10		21	25	29
94	—	—	1	9 − 10	9 − 10		9 − 10		21	25	29
95	—	—	2	9 − 10	9 − 10		9 − 10		21	25	30
96	—	—	2	9 − 10	9 − 10		9 − 10		21	25	30
97	—	—	2	9 − 11	9 − 10		9 − 10		22	25	30
98	—	—	2	9 − 11	9 − 10		9 − 10		22	25	30
99	—	—	2	9 − 11	9 − 10		9 − 10		22	26	30
100	—	0	2	9 − 11	9 − 11		9 − 11		22	26	31
200	3	6	8	19 − 21	19 − 21		19 − 21		35	40	46
300	9	12	16	29 − 31	29 − 31		29 − 31		48	53	60
400	15	19	23	39 − 41	39 − 41		39 − 41		60	66	74
500	22	27	31	49 − 51	49 − 51		49 − 51		72	79	87
750	40	46	52	74 − 76	74 − 76		74 − 76		101	109	119
1000	59	67	74	99 − 101	99 − 101		99 − 101		129	139	150
1500	99	109	118	148 − 152	149 − 151		149 − 151		185	197	210
2000	140	152	163	198 − 202	199 − 201		199 − 201		240	253	268

[a] See text for nomenclature.

Table 6
VALUES OF REGRESSION EQUATION CONSTANTS REQUIRED
FOR CALCULATING M, M_o, M', and M''

Probability	Statistical significance p	M		M_o	
		d	e	d_o	e_o
1/2	0.05	0.20559	−0.51831	—	—
	0.01	0.28672	−0.50629	—	—
	0.001	0.36307	−0.49685	—	—
1/3	0.05	0.24795	−0.53893	+0.12870	−0.50308
	0.01	0.34584	−0.53192	+0.20873	−0.49182
	0.001	0.43211	−0.52477	+0.27593	−0.48030
1/5	0.05	0.26073	−0.56433	+0.016254	−0.49130
	0.01	0.37432	−0.56197	+0.077633	−0.47476
	0.001	0.47820	−0.55991	+0.12514	−0.45770
1/7.5	0.05	0.26233	−0.58600	−0.084849	−0.48253
	0.01	0.38676	−0.58680	−0.039561	−0.46146
	0.001	0.50285	−0.58817	−0.0085423	−0.43979
1/10	0.05	0.26388	−0.60264	−0.16152	−0.47607
	0.01	0.39555	−0.60549	−0.12879	−0.45147
	0.001	0.51967	−0.60905	−0.11034	−0.42630
1/2	0.95	−0.21391	−0.80357	−0.21391	−0.80357
	0.99	−0.078200	−0.93860	−0.078200	−0.93860
	0.999	−0.0094739	−0.99287	−0.0094739	−0.99287
1/3	0.95	−0.20683	−0.81074	−0.21103	−0.80945
	0.99	−0.074145	−0.94173	−0.074984	−0.94148
	0.999	−0.0089027	−0.99329	−0.0089865	−0.99326
1/5	0.95	−0.18999	−0.89219	−0.19890	−0.82646
	0.99	−0.064230	−0.94961	−0.066010	−0.94907
	0.999	−0.0075283	−0.99430	−0.0077063	−0.99425
1/7.5	0.95	−0.17315	−0.84656	−0.18598	−0.84263
	0.99	−0.055468	−0.95650	−0.058032	−0.95572
	0.999	−0.0063668	−0.99516	−0.0066232	−0.99508
1/10	0.95	−0.16010	−0.85929	−0.17599	−0.85442
	0.99	−0.049375	−0.96124	−0.052551	−0.96026
	0.999	−0.0055879	−0.99574	−0.0059055	−0.99564

Note: d_i = constant terms, e_i = factor of variable terms.

If the values of P'_a and P'_b are fractional, they must be rounded towards each other to the next integral values before entering significance tables.

APPENDIX II:

GENERAL DATA CORRECTION FOLLOWING A DIFFERENCE TEST

N_t = total number of assessors
N_c = number of assessors who make a correct identification
$1/B$ = probability of identifying the product correctly by chance

where $B = {}^nC_r$ when $r_1 \neq r_2$ or $B = 0.5\ {}^nC_r$ when $r_1 = r_2$
in a polygonal test with n samples of which r are of one product,
and $B = {}^nC_r$ in a polyhedral test with n samples of which r are of one particular product,

N_g = number of "lucky guessers" = $(N_t - N_c) / (B - 1)$

(i)   In a polygonal test,

P_a = number of assessors, of N_c, who prefer product A

P_b = number of assessors, of N_c, who prefered product B

then the corrected numbers of preferences are

$P'_a = P_a - 0.5 N_g$

and $P'_b = P_b - 0.5 N_g$

of $N'_c = N_c - N_g = P'_a + P'_b$

In a polyhedral test,

T_q = preference rank total for product Q, from N_c assessors

F = random rank = $0.5 \times (1 + \text{number of products ranked})$

then the corrected preference rank total is

$T'_q = T_q - F N_g$ for the predetermined product Q among

$N'_c = N_c - N_g$

When N'_c is fractional, it must be rounded downward to the nearest integer for determining most preferred significance or rounded upwards for determining least preferred significance.

(ii)   S = mean score of a product Q, of N_c (on, say, a 0-to-10 scale), with variance = s^2

S_i = mean score of incorrectly identified samples, with variance = s^2_i

then $S N_c = S_i N_g + S' N'_c$

which can be solved for S', the corrected mean score.

The variance s^2 may be considered to have arisen from the pooling[16] of the corrected variance $(s')^2$ and the "guessers" variance s^2_i:

$(N_c - 2) s^2 = (N'_c - 1) (s')^2 + (N_g - 1) s^2_i$

which can be solved for $(s')^2$.

If N_c is small, invalid results may be obtained, which are exemplified by finding S' outside the scale limits or by a negative value for $(s')^2$.

(iii)   At each nonparametric rating level for product Q, R_c = number of ratings due to correctly identified samples

R_i = number of ratings due to incorrectly identified samples

while ΣR_i = total R_i for that product, then the corrected number of ratings is $R'_c = R_c - (R_i N_g / \Sigma R_i)$

When R_c is zero (or very small) and R_i is not zero, R'_c would be unacceptably negative; the apparent problem disappears if adjacent values in the rating scale are combined to increase the size of R_c.

APPENDIX III
CHI-SQUARE MODEL

General Contingency Table

	Distinguished	Not distinguished	Totals
Test result	F N	N − F N	N
Null hypothesis	N/B	N − N/B	N
Totals	N(F + 1/B)	N(2 − F − 1/B)	2N

where

F = fraction distinguished (when B > 2) or preferred (when B = 2)

N = total number of assessors

1/B = probability of obtaining result by chance

Critical values of F are calculated from critical values (H) of χ^2, allowing for the "correction for continuity" to 2 × 2 contingency tables:[19]

$$a\,F^2 + b\,F + c = 0 \tag{1}$$

where

a = $B^2\,N(2N + H)$

b = $-2B\,N\{2(N + B) + (B - 1)H\}$

c = $2(N + B)^2 - (2B - 1)N\,H$

(i)   Using the positive square root in the solution to Equation 1,

$$M \geqslant F\,N$$

For large values of N, calculate the plurality (P):[20]

$$P = F - 1/B$$

and then

$$\log P = d + e \log N$$

is practically linear (correlation coefficient $> |0.999|$). Because the relationship is not *exactly* linear, slightly different values of d and e are found depending on the precise values of P and N correlated; the minimum values of M derived here are thus not absolute, but constitute sufficiently close approximations for all practical purposes.

(ii)   With critical values of χ^2 close to zero, using the positive square root in the solution in Equation 1 gives

$$M' \leqslant F\,N$$

i.e., $M' \leqslant N(1/B + P)$ where $\log P = d' + e' \log N$.

Using the negative square root in the solution in Equation 1, putting

$$b'' = -2B\,N[2(N - B) + (B - 1)H]$$

$$c'' = 2(N - B)^2 - (2B - 1)N\,H$$

and because P is now negative, replacing it with

$$\bar{P} = -P$$

gives $M'' \geqslant N(1/B - \bar{P})$ where $\log \bar{P} = d'' + e'' \log N$

(iii)   Returning to high critical values of χ^2, and using the negative square root in the solution to Equation 1,

$$M_0 \geqslant N(1/B - \bar{P})$$

where $\log \bar{P} = d_0 + e_0 \log N$.

REFERENCES

1. British Standard B.S. 5098, *Glossary of Terms Relating to Sensory Analysis of Food,* British Standards Institution, London, 1975.
2. ASTM Committee E-18 on Sensory Evaluation of Materials and Products, *Manual on Sensory Testing Methods,* ASTM Special Technical Publication 434, American Society for Testing and Materials, Philadelphia, 1968.
3. **Helm, E. and Trolle, B.,** Selection of a taste panel, *Wallerstein Lab. Commun.,* 9, 181, 1946.
4. **Wittes, J. and Turk, A.,** The selection of judges for odor discrimination panels, in *Correlation of Subjective-Objective Methods in the Study of Odors and Taste,* Stahl, W. H., Chairman, ASTM Special Technical Publication 440, American Society for Testing and Materials, Philadelphia, 1968, 49.
5. **Basker, D.,** Comparison of discrimination ability between taste panel assessors, *Chem. Senses Flavor,* 2, 207, 1976.
6. **Basker, D.,** The number of assessors required for taste panels, *Chem. Senses Flavor,* 2, 493, 1977.
7. **Woodward, W. A. and Schucany, W. R.,** Combination of a preference pattern with the triangle taste test, *Biometrics,* 33, 31, 1977.
8. **Schutz, H. G. and Bradley, J.,** Effect of bias on preference in the difference-preference test, in *Food Acceptance Tasting Methodology,* Peryam, D., Pilgrim, F. J., and Peterson, M. S., Eds., Quartermaster Food and Container Institute, Chicago, 1954, 85.
9. **Basker, D.,** Effect of selection on the ratings of taste panel assessors, *J. Food Technol.,* 12, 599, 1977.
10. **Moskowitz, H. R.,** *Product Testing and Sensory Evaluation of Foods,* Food & Nutrition Press, Westport, Conn., 1983.
11. **Basker, D.,** Polygonal and polyhedral taste testing, *J. Food Qual.,* 3, 1, 1980.
12. **Basker, D.,** Further polyhedral taste panel difference tests, *J. Food Qual.,* 4, 229, 1981.
13. **Kramer, A. and Twigg, B. A.,** *Fundamentals of Quality Control for the Food Industry,* AVI Publ., Westport, Conn., 1962, 114.
14. **Potter, N. N.,** *Food Science,* AVI Publ., Westport, Conn., 1968, 115.
15. **Kahan, G., Cooper, D., Papavasilio, A., and Kramer, A.,** Expanded tables for determining significance of differences for ranked data, *Food Technol.,* 27(5), 63, 1973.
16. **Dixon, W. J. and Massey, F. J.,** *Introduction to Statistical Analysis,* McGraw-Hill, New York, 1969, 238.
17. **Basker, D.,** Nonparametric comparison of samples by taste panels: limits of difference detectable and establishment of nondifference, *J. Food Qual.,* 5, 1, 1981.
18. **de Cervantes, M.,** *Don Quixote,* Book IV (Part I), 1606, chap. 3.
19. **Duncan, A. J.,** *Quality Control and Industrial Statistics,* 3rd ed., Richard D. Irwin, Homewood, Ill., 1965, 536.
20. **Basker, D.,** The number of assessors required for nonparametric taste difference tests, *J. Food Qual.,* 4, 101, 1980.

Chapter 8

SENSORY DIFFERENCE AND PREFERENCE TESTING: THE USE OF SIGNAL DETECTION MEASURES

Michael O'Mahony

TABLE OF CONTENTS

I.   Introduction 146

II.   Sensory Evaluation and Sensory Difference Testing 146
        A.   The Applications of Sensory Testing 146
        B.   Goals and Strategies 147
                1.   Sensory Evaluation I 147
                2.   Sensory Evaluation II 148
                3.   Consumer Testing 149
                4.   Sensory Psychophysics 150
        C.   Sensory Difference Tests 150
                1.   The Need for Forced-Choice Procedures 150
                2.   A Summary of Testing Procedures 151
        D.   Threshold and Preference Tests 152
        E.   The Problem of Multiple Difference Testing 153

III.   A Brief Summary of Classical Signal Detection Theory 153
        A.   Signals and Noise 153
        B.   d′: A Measure of Signal Strength 154
        C.   Measuring d′ and P(A): ROC Curves 155
                1.   Constructing an ROC Curve 155
                2.   P(A): The Proportion of Area Under an ROC Curve 156
                3.   Can We Now Forget About d′ and P(A)? 156

IV.   The Use of d′ in Sensory Difference Testing 156

V.   Computation of the R-Index 157
        A.   The R-Index by Rating: Two Food Treatments 157
        B.   The R-Index as a Threshold Measure 160
        C.   The R-Index by Rating: More Than Two Food Treatments 160
        D.   The R-Index is not a Substitute for Intensity Scaling 162

VI.   Statistical Treatment of R-Index Values 162
        A.   Analysis of Variance 162
        B.   Rank Sums Test 163
        C.   Statistical Analysis and the Design of Sensory Analytic Tests 165

VII.   Application to Consumer Preference Testing 166

VIII.   Summary 167

IX.   Appendix: Computer Program for Computation and Statistical
        Analysis of the R-Index 167

References 174

I. INTRODUCTION

Before proceeding with an explanation of signal detection theory and its use in the sensory evaluation of food, it is worthwhile first to consider the uses and goals of sensory evaluation and the role of difference testing. Along with this, psychophysics and consumer testing will be considered with special reference to threshold and preference testing. This is necessary because the exact goals and strategies of sensory evaluation are often confused, and it is only against a background of its restated logic that the uses and potential of signal detection can fully be understood.

Having set the scene, classical signal detection theory will be outlined briefly. Here, the discussion will be mainly psychophysical because it was for psychophysics that the approach was developed. It will become apparent that in its classical form, signal detection has little application in the sensory evaluation in food. The main reason for the brief review of signal detection is to allow the reader to understand that the far simpler measures discussed for difference testing are in fact signal detection measures.

New approaches and adaptations of signal detection to difference testing will then be considered. These are both conceptually simpler and shorter than the classical methods and are thus easily applicable to sensory evaluation. Their application will be discussed, bearing in mind the varying aims of sensory testing.

II. SENSORY EVALUATION AND SENSORY DIFFERENCE TESTING

A. The Applications of Sensory Testing

The use of the human senses to measure the flavor and sensory characteristics of foods and other products is called Sensory Analysis or Sensory Evaluation. Human senses are used because they cannot as yet be reproduced exactly by laboratory instruments. Human senses and laboratory instruments do not have the same sensitivity; a human nose can detect volatile chemicals that are not detected by a gas chromatograph. The human senses also integrate their input so that changes in one sense can be perceived as changes in another. A change in smell can sometimes be perceived as a change in taste; this is a feature that is sensibly avoided by manufacturers of laboratory instrumentation.

Sensory evaluation can be a useful tool in the development of new products and new food resources, in storage and shelf-life studies, and in quality assurance of products, especially with changes of formulation, processing, and packaging. It is used for the grading and pricing of products and in more basic research to determine exactly what chemical changes in a food affect its flavor. In the latter case, a full understanding of, say, the chemical changes taking place over the aging period of a vintage red wine could allow intervention so as to speed up the process without the production of off-flavors.

A related area is that of consumer testing. While sensory evaluation is used to determine "What is the flavor of the food?", consumer testing is used to determine "Do the customers like the food? Will they buy it?" Another related area is psychophysics. A sensory psychophysicist is a basic researcher who uses behavioral rather than physiological techniques to study the working of the senses (generally in humans). Such research is used to gain knowledge of the functioning of the brain and the senses. Besides telling us more about human behavior, such information is useful to the communications, artificial intelligence, computing, and robotics industries. The nervous system is a sophisiticated communications system that transmits massive amounts of information over comparatively few channels or "wires"; the communications industry could learn much from such a system. The robots of the future will need to see, sense, and process massive amounts of information, a task done with ease by the human sensory system. Unlocking the secrets of the human senses can provide breakthroughs for these industries.

There are medical reasons for studying the senses. First, it would be desirable to be able to repair and replace defective parts of the human senses when they fail. Changes in sensory functioning can also be indicative of disease, genetic variation, or bodily malfunctions. Transduction mechanisms of taste, the reactions of taste molecules with the membrane of a taste receptor, are an example of a process called chemoreception. The study of easily accessible chemoreception mechanisms such as taste and smell can provide useful information for less accessible chemoreception mechanisms such as the action of drugs and hormones or the working of some brain and cell mechanisms. Also related here are studies of the chemical sensing of pheromones, attractants, and repellants for both animals and insects. Sensory psychophysics is also used in many diverse areas ranging from the study of nutrition and food habits to the design of instrument panels.

The uses of sensory measurement are diverse, encompassing a range of disciplines of which food science forms only one part. However, although the aims of the research may be very different, the methods are often similar. This can be an advantage, allowing economy of method. Yet, it can also cause confusion. Unfortunately, with the sensory evaluation of foods, there has been considerable confusion caused by the unclear definition of experimental goals.

B. Goals and Strategies
1. Sensory Evaluation I
There are at least two completely different types of sensory evaluation. Here they will be called Sensory Evaluation I and Sensory Evaluation II.

Sensory Evaluation I refers to sensory evaluation when it is used as a tool to study chemical and physical characteristics of food systems. It can be used to determine, say, whether a change in processing alters the acid content of a food. This would be detected by human judges as a change in sourness. A change in volatile chemical content would be detected as a change in odor. A change in color may indicate chemical changes such as browning reactions, etc. Here the focus of the research is on the food. The human senses are merely a broad-purpose instrument, an array of sensors connected to a powerful computer, which can be used to measure various properties of that food.

If the aim is to see how a new processing treatment affects the food, the human judge can be used to determine in what area any changes have taken place. He can be used for a general screening analysis to be supplemented by further instrumental measures, or he can even be used as a substitute where instrumental analysis is inconvenient or insufficiently sensitive. Here, a sample of food portions are assessed by the judge and statistical inferences made about the population from which these portions of food were sampled. The "N" in the statistical analysis (binomial test) refers to the number of replicate portions tasted by a given judge; data for each judge is analyzed separately. As with instrumental analysis, one good instrument, the most sensitive available, is all that is required. Yet, we do not know the workings or characteristics of a human judge in the way that we understand laboratory instruments; there is no handbook or wiring diagram supplied with a human judge. Thus, more than one judge is used as a fail-safe; if some judges do not perform as well, we use the data from those that do. In this way, the best human instrument or instruments for a specific task are screened and attention is paid to the data they provide on the samples of food.

For such an analysis, every precaution is taken to assure that the human judge is working with maximum sensitivity and reliability. For visual judgments, the food is not tasted so as to avoid interference from taste, smell, or texture cues. Food can be mashed, ground, or juiced to avoid bias due to texture differences. Noseclips can be worn so as to avoid smelling a food. A comparison of, say, the sourness of two oranges can involve using juice to avoid texture and appearance cues, use of colored light to mask color differences, use of a noseclip to prevent detection of odor differences, and use of a prerinse of gymnema sylvestre tea[1,2] to block the sweet taste receptors. In this way, other sensory information is eliminated or minimized to allow the judge to concentrate on sourness. It is even feasible for liquid food to be flowed over the tongue rather than sipped in the normal way. Such a technique is used in psychophysics to increase taste sensitivity because it controls changes caused by sensory adaptation (eliminates "taste fatigue") and allows greater precision of measurement.[3]

Although these techniques remove the human from the normal eating situation, their high degree of control is essential for establishing where sensory differences exist between different samples of food. Accordingly, the human judge is a sensitive multipurpose instrument for research and development. Such an instrument is extremely useful when reprocessing or reformulation are to be manipulated to minimize flavor loss or the development of undesirable sensory characteristics.

2. Sensory Evaluation II

There is still a further question, and this is where Sensory Evaluation II comes in. Having established where the off-flavors or sensory changes have occurred in a food, there comes a point where it may no longer be worth manipulating the processing or reformulation to reduce these changes any further. They may be so small that they are not noticed except under highly controlled conditions by the most sensitive judges. The ordinary consumer may never notice them. The second question then, becomes one of whether ordinary consumers under ordinary eating conditions would notice the changes. It is here that sensory evaluation has a second use, which here will be called Sensory Evaluation II.

For Sensory Evaluation II, a sample or panel of typical consumers must try to determine whether they can distinguish differences between the different foods. Here, selection of judges is not limited to the most sensitive, nor are they given any special training to make them more familiar or more skilled at perceiving the particular nuances of the food flavor, as with Sensory Evaluation I. If this were done, they would no longer by typical of the consumers. Accordingly, judges are chosen who are as close as possible to the consuming population. A panel selected from the work place should usually be sufficient. Samples need not be taken from all around the country as with consumer testing because regional differences are not expected in the flavor sensitivity of our judges in the way that regional differences exist in food preference. Of course, should the possibility of geographical differences be suspected in the sensory physiology of the consumers (perhaps because of racial variation), it is worth controlling by selecting appropriate panelists. Generally, however, it is more likely that sensitivity to flavor changes would not vary over regions of the country in the way that food preferences do.

The judges can, of course, be trained in sensory methodology. General practice or familiarity with scaling or difference testing is not expected to alter the judges' familiarity with the food under question. As long as the judge does not practice with the test food, he should not become more familiar with it than the typical consumer. Even so, it is probably advisable to keep practice to a minimum. Practice on one food may alter how a judge will attend to another food, making him no longer typical. It is thus better to use behavioral methods which require little or no practice. Simple difference tests or signal detection procedures can be understood with little or no practice.

Here, an attempt is being made to reproduce normal eating conditions. A judge would not be asked to concentrate only on smell or only on texture unless he would do this normally. Certainly, the senses are not isolated with noseclips and colored lights as they would be for Sensory Evaluation I. A simple triangle test with instructions to eat the food in the normal way is quite sufficient. It is likely that although differences may be detected by screened panelists in controlled conditions for Sensory Evaluation I, they may not be detected by an unscreened "typical" panel under ordinary eating conditions for Sensory Evaluation II. In this case, it may be desirable to go ahead and market the reformulated or reprocessed product. Essentially, Sensory Evaluation II gives us a measure of whether consumers would detect changes. From this, a marketing decision can be made about whether to go ahead and risk the product in the marketplace or whether to reformulate further.

It is important to remember here that the panel of judges now constitutes a sample from which inferences are made about the population. They are no longer instruments for testing the samples of food. The food portions, whose characteristics are known from Sensory Evaluation I, are used to test samples of people. The "N" in the statistical analysis (binomial test) is now the number of judges.

Should the sample of typical judges notice differences in the food, further manipulation of the food may be necessary. These judges may indicate where they think the differences occur (color, taste, smell, etc.) but is should be remembered that they are untrained and open to cross-sensory bias (for example, they may mistake changes in smell for changes in taste). They may be useful for pointing out obvious large differences, but then such tests would hardly be performed if the differences were that obvious. The trained panel under controlled conditions is the proper source of information about the sensory characteristics of the food.

3. Consumer Testing

If the differences in sensory characteristics cannot be eliminated, the marketer may just have to go ahead and market the food anyway. Perhaps the flavor of a rival company's brand leader could not be completely mimicked, or perhaps a new processing or packaging method brought about flavor changes that could not be prevented. Some consumer testing would then have to be done to determine whether typical consumers will accept or buy the food. This section will refer to that aspect of consumer testing that examines consumer food preferences.

Because food preferences vary around the country, samples of typical consumers would need to be chosen from around the country. This is an expensive operation. A considerable saving in cost is thus possible if sufficient data can be obtained merely by sensory evaluation (e.g., the ordinary consumer cannot tell a difference under normal eating conditions). As with Sensory Evaluation II, consumer testing requires that the food should be tasted under normal eating conditions. The data from the sample of consumers are used to gain an estimate of preferences in the population. As with Sensory Evaluation II, the "N" for the statistics (binomial test) is the number of consumers in the sample. The methods should be kept simple because a consumer generally has little time to learn complex techniques such as scaling or accurate description of off-flavors. His data should be treated as coming from someone who is unskilled in scaling or descriptive analysis. This makes nonsense of much consumer research which requires consumers to generate vast amounts of reliable scaled descriptive data, which are then subjected to complex parametric statistical analysis. The numbers that consumers generate are very likely non-numeric, while their sensory descriptions are unstandardized. An analysis of such data would be at best "imprecise" and at worst "garbage". A simple approach such as paired preference testing or ranking would seem more likely to generate sensible data.

4. Sensory Psychophysics

Finally, there is sensory psychophysics, the branch of psychology devoted to finding out how the human sensory system works. Samples of judges are examined and inferences made about the population, as in Sensory Evaluation II. The "N" in the statistical analysis is the number of judges (or judges × experimental treatments). Judges are placed in various experimental conditions and their changes in performance used to determine how the senses work. Whereas the sensory analyst is interested in difference testing, the psychophysicist is interested in judge sensitivity and how it may change. The sensory analyst uses scaling to attribute numerical values to the strength of various sensory characteristics of a food; the psychophysicist uses scaling to understand the relationship between physical intensity and perceived intensity of a stimulus. Whereas a sensory analyst uses a standardized language to describe flavor characteristics of a food, a psychophysicist (or a psycholinguist) is often interested in the nature of everyday language, and the development and processing of sensory concepts. However, for precise measurements of taste or smell per se, the psychophysicist should use a standardized language. Some of the methods in sensory analysis and psychophysics may be the same, but the goals are different.

The point about the goals of sensory evaluation, psychophysics, and consumer testing has been labored, yet it is essential that these are clear to avoid choosing an inappropriate method or using an incorrect statistical analysis. It is also important to understand the often confused yet quite separate goals and approaches of Sensory Evaluation I and II. It is only when these are clear that an appreciation of the use of signal detection is possible.

C. Sensory Difference Tests

1. The Need for Forced-Choice Procedures

Essentially, sensory difference tests are designed to determine whether a judge can or cannot detect any differences in the sensory characteristics of foods or other products. The tests must solve two main problems. First, they must require the judge not merely to say whether he can distinguish between two foods; a judge's word in such matters cannot be trusted. The judge must demonstrate that he can do so. This can be done by requiring him to select an odd sample from a group or identify a sample that is the same as a standard sample. Second, difference tests must solve the problem of the implicit question: "How different do two foods have to be before they can be called different?" This may not seem like an important question, but the effects of "where the line is drawn" are profound. To warrant difference testing, any differences will be slight. They will often be so slight that the judge is unsure whether or not a difference even exists. A reckless judge may go ahead and state that there is a difference, even when unsure. In a more cautious mood, the same judge would not have committed himself to saying a difference existed, yet, the sensitivity of his taste or smell receptors would have been the same. His response would have varied merely because of his particular degree of cautiousness at that time. Judges "draw the line" in a different place according to their caution. The difference must be greater for the more cautious judge to report a difference. In signal detection parlance[4] the "criterion" of difference must be greater for him to report a difference. Obviously, one is interested in how well a judge can detect differences, not the variation of his criterion.

Uncontrolled changes in the criterion can be eliminated by using a "forced-choice" procedure. This avoids the necessity of the judge ever having to "draw the line." It is drawn for him by saying that the food samples in a test are sufficiently different for them to be reported as different; the judge merely has to indicate which sample is the odd one. The forced choice procedure is not the only way of eliminating criterion problems; the signal detection methods outlined later do it in a different way.

2. A Summary of Testing Procedures

A variety of difference testing procedures are available.[5] All involve a forced choice procedure to avert criterion problems, and all require the judge to demonstrate rather than merely state that he can distinguish between two foods. The various procedures vary in their sensitivity for a variety of reasons that are beyond the scope of this article.[6]

The paired comparison procedure requires the judge to distinguish which of two samples has a greater degree of a given attribute such as sweetness, firmness of texture, darkness of color, etc. The judge knows that one sample will be darker, sweeter, or whatever, so he does not have to worry about whether the two samples are sufficiently different to be reported as different. If the judge consistently picks the target sample over replicate testings, he can tell the difference between the two; if he does not, he cannot tell the difference. The target sample may be the one with the greater amount of the attribute. If it is not possible to know in advance which sample will have the greater amount of the attribute, then consistency of response is taken as an indication that a difference has been detected.

Sometimes it may be difficult to convey in words the nature of the difference, if any, between the sample. A change in processing may bring about subtle changes in flavor for which there are no words to describe. In this case, a triangle test may be used. The task here is to pick out the odd sample from a set of three (two of one treatment, one of the other). Consistent selection of the odd sample from the other two in replicate tests indicates that the judge can differentiate between the two food treatments. Traditionally, for the triangle test, the odd sample is varied between the two food treatments and the instructions are to pick the odd sample. A version of the triangle test exists whereby the odd sample always comes from the same food treatment and the instructions specify the nature of the difference (pick the sweeter sample). This is called the three-alternative forced choice, or 3-AFC test.[4]

Although the triangular design of picking one odd sample from three is commonly used, other variants are possible. Yet such variants are rarely used and seldom studied, probably because of an unjustified assumption[7] that tests that use more samples are necessarily less sensitive to differences. One variant that has been used is to pick four from a set of eight (four of one treatment, four of another). This procedure has been called an octad test[8] but is more generally called the Harris-Kalmus test.[9] Under the latter name, it has been commonly used to determine whether judges can distinguish between a solution of a given concentration of phenylthiocarbamide (PTC, also called phenylthiourea) and water; it is used to determine PTC thresholds for genetic research. A further variant, the Fallis-Lasagna-Tétreault test,[10] has also been used for threshold measurement.[11,12,13] Here, one odd sample is picked from four (one taste solution, three water samples).

Another approach to difference testing is to require a judge to identify from two "test" samples the one that is the same as a previously presented "standard" sample. This is the commonly used duo-trio test. Should the judge be able to do this consistently over replicate tests, it is taken as evidence that he can distinguish between the two test samples. This test differs from the triangle test in that the judge is searching for similarities rather than differences. The cognitive effect of such a different search procedure is little understood; however, the criterion problem is solved because the judge is instructed that the two test samples are different but one will be the same as the standard. The judge does not have to busy himself with the question of whether one of the test samples is sufficiently similar to the standard to be reported as the same.

A less popular variant is the dual standard test whereby two standards are first presented, one the same as one test sample, one the same as the other. The judge then has to match the two unknown samples to the two standards. This test can be more sensitive than the duo-trio.[7] Another variant is the tetrade test, whereby the judge has to select from three samples (two of one treatment, one of the other) the one or two samples that are the same as a previously presented standard.[14]

These tests sensibly use a forced-choice protocol to circumvent criterion variation. For determining whether a given judge can tell the difference between the two food treatments consistently over replicate tests, binomial statistics are appropriate.[15,16] For Sensory Evaluation I, the most sensitive judges would perform many replicate tests to obtain a good sample of food portions from both treatments. For Sensory Evaluation II, a sample of judges is examined so as to infer the proportion in the population who can determine the difference. Again, binomial statistics are appropriate for determining whether a significant majority can or cannot tell the difference.

D. Threshold and Preference Tests

A threshold test is designed to determine the minimum concentration at which a stimulus can be detected as different from a "blank" (detection threshold) or can be recognized as having its typical taste or smell (recognition threshold). There are a variety of procedures for determining thresholds.[17] Many suffer from difficulties with criterion variation.

Many detection threshold procedures require a judge to taste a succession of samples and judge whether each one is a stimulus or a "blank". For taste, the stimulus is typically a given concentration of an aqueous solution, while the "blank" is a water sample. For smell, the stimulus will be a container of a given concentration of volatiles and the blank an empty container.

Unfortunately, such procedures require the judge to answer the implicit criterion question for each sample: "Is the sample sufficiently different from the "blank" to be judged as a stimulus?" The problem can be solved by incorporating a forced choice procedure into the threshold test. Consistently correct responses would indicate that this concentration of stimulus can be differentiated from water. Such a test is really a difference test, because the samples could be differentiated not only because of the particular taste of the stimulus solutions but also because of the watery taste of the water samples. The same can be true for smell depending on the adaptation state of the smell receptors.

A recognition threshold, by definition, involves criterion variation and is best avoided. It requires a judge to determine whether a stimulus is sufficiently strong to be judged as having a typical taste or smell. The judgment implies the criterion question regarding how strong a stimulus taste or smell has to be, to be regarded as "typical". There is great potential for variation here, as can be seen from the decisions taken on seemingly simple judgments such as saltiness[18] or sourness.[19,20] No presentation procedure can get around this problem. Even a forced-choice procedure requires the judge to decide whether the stimulus is sufficiently typical to be worthy of being picked. Methods which avoid such a decision become detection threshold measures.

Reluctantly, one has to conclude that only detection thresholds incorporating a forced choice avoid criterion difficulties, and these are essentially difference tests. Other detection procedures and all recognition procedures have inherent criterion problems and should be avoided. One tolerable circumstance may be a within-subjects comparison with experimental conditions that produce differences in measured threshold that are sufficiently large to render criterion variation of minor importance.

Preference tests are somewhat similar to difference tests. They are designed to determine which of two products are preferred by a judge. Here the judges are generally consumers sampled to be typical of potential customers for a given product. Generally, a paired comparison procedure is used, the judge indicating which of two products he prefers. A triangular or duo-trio procedure would hardly be appropriate because it would involve confusing instructions (e.g., pick the one or the two samples you prefer from the three presented). The paired comparison may force the judge to pick a preferred sample or the procedure may allow a "no preference" response. This will depend on which risk the marketing division wishes to take. They could risk forcing a choice of product, when the two products are sufficiently liked that both might be purchased. In taking this risk, it is assumed that equal numbers of "no preference" consumers will pick each product. They could decide not to take this risk but instead risk the effects of criterion variation by allowing a "no preference" category; this introduces the question of how strong a preference has to be before it is reported as a definite preference. Yet, consumer researchers may be prepared to take this risk so as to get an estimate of the number of consumers who do not have a strong preference. A combination test, in which "no preference" was allowed, followed by a procedure that forced a preference, could gather the information in both ways.

Binomial statistics can be used to test for the consistency of response of a given judge or the determination of whether a majority response in a group of judges is significant.

E. The Problem of Multiple Difference Testing

A difference test determines whether a judge can distinguish between two food treatments by whatever sensory characteristic is under consideration. Binomial statistics give a "yes" or "no" answer at a given level of significance. However, it is quite common for some measure of degree of difference to be required rather than merely a dichotomous "difference" or "no difference" result. It may be that several formulations are being tested against a market leader and a measure is required of the degree of difference between each formulation and that market leader. The general approach is to measure the degree of difference by a scaling method, of which there are many, ranging from category and graphic scales to magnitude estimation. However, all of these procedures were designed to determine the degree of difference between stimuli that were obviously different. The evidence available would suggest that humans are not good number generators.[21] Any numerical value obtained from a scale would be expected to be fairly approximate, yet the error involved would not be too important for stimuli that are obviously different (3 ± 2 is different from 8 ± 2).

However, if scaling is performed after a difference test, the stimuli would be very similar, or else difference tests would not have been necessary. In this case, it is very likely that the error involved in scaling would be greater than the differences being measured (e.g., 2 ± 2 could not be distinguished from 3.5 ± 2). Scaling would seem insufficiently precise for the measurement being made. What is needed is a measure of degree of difference, obtained without the judge ever having to generate numbers himself. Signal detection procedures provide this. They provide a series of simple behavioral measures which avoid the necessity for the judge to use any form of numerical scaling procedure. Yet, numerical measures of the degree of difference between foods can be computed straightforwardly from the judge's simple responses. Furthermore, the measures obtained are probability measures which are susceptible to parametric statistical analysis; such analysis for scaled data is arguable.[21]

III. A BRIEF SUMMARY OF CLASSICAL SIGNAL DETECTION THEORY

A. Signals and Noise

Signal detection is discussed briefly here to give sufficient background for the reader to be able to understand how the simple measures for sensory difference testing, detailed in Section IV, relate to classical signal detection theory. The reader who is not interested can skip to Section IV.

Signal detection theory[4] treats the sensory system like a communications system in which the higher centers of the brain are seen as the recipients of input. This input can be the random spontaneous firing of the nervous system called ''noise''. On the other hand, ''signals'' sent from the sensory organs also constitute input and these must be distinguished from the background noise. If the signal is large, it is easy for the brain to distinguish it from the noise. The noise level may be varying continuously but this variation will not be mistaken for the relatively large input that constitutes most signals from the senses. Of course, the brain does not know that the increase in input is necessarily a signal. It sets a criterion level of input, and anything that exceeds this level is deemed to be a signal. This criterion level is arbitrary. If the judge is being cautious about saying there is a signal coming in when there might not be, he sets a high criterion level. This situation could come about by the judge being punished for such false alarms. On the other hand, should the judge be encouraged by some reward system in the experiment to report signals, he will choose a low level of input as his criterion. This criterion will be exceeded more often and more signals will be reported.

The variation of this criterion level of input is not to be confused with the sensitivity of the sensory system. A more sensitive sensory system is one that will respond with a bigger signal for a given stimulus. The criterion is merely a matter of where the judge ''draws the line'' between reporting a signal or mere noise. A given level of input could be taken as noise or as a signal depending on the criterion.

For large signals this poses no problem. The massive increase in input that constitutes most signals is clearly recognizable. The realm of threshold measurement is where it becomes a problem. It is precisely here that the signal is minimal. It is so small that it is easily confused with noise. Changes in the criterion level can have a profound effect on whether a small increase in input is reported as a signal or as a noise.

As mentioned in Section II.D, a threshold procedure that requires a judge to rate a single stimulus as either signal (a given chemical taste solution) or noise (a water ''blank'') is not a direct measure of sensitivity per se; it is a measure of sensitivity modified by where the judge sets his criterion. The need here is for a procedure that provides a measure of the level of input associated with a given stimulus, regardless of the criterion. Signal detection provides such a measure.

B. d': A Measure of Signal Strength

The forced-choice procedure detailed in Section II.C.1, is a means of setting the criterion level between two input levels. It is known that one will be a signal and the other a noise. The judge has merely to say which is which. This is a convenient procedure for setting the criterion at an appropriate level for the particular task at hand.

The signal detection approach is to get a measure of the signal strength independent of any variation in the criterion level. This measure is called d' (d-prime). The first problem is to choose a unit for the measure of signal strength. Physiological measures would be impractical, so the signal is measured in terms of the variation of the noise level.

The noise level is not constant; it rises and falls continually. Sometimes it is high, and sometimes it is low, but it will vary about a mean level. The first assumption is that this variation in noise level over time can be described by a normal distribution. It is the standard deviation of this distribution which is the unit chosen for d', the signal strength. Thus, a signal strength of $d' = 3$ means that the signal increases the input level to the brain by an amount three times the standard deviation of the variation of the noise level. It is a ''signal to noise ratio''. Such a measure is useful because it describes the signal strength in terms of each individual subject's own variation in noise level. It takes into account the fact that a subject with a larger variation in noise level will need a larger increase in input before he distinguishes it from mere noise variation and decides it is a signal.

C. Measuring d′ and P(A): ROC Curves

1. Constructing an ROC Curve

The exact details of the computation of d′ using the classical signal detection methods are beyond the scope of this article, but it is outlined in detail elsewhere.[4,22] Here, it will be described only briefly.

Traditionally, d′ is calculated by constructing what is called a Receiver Operating Characteristic (ROC) curve. d′ can be calculated from the degree of curvature of this curve.

The ROC curve is plotted in the following way. The responses to signals and noises are noted, in particular, the proportion of times a judge says a signal is present when it actually is (proportion of "hits") and the proportion of times he says a signal is present when it is not (proportion of false alarms). The proportion of hits and false alarms is noted and this provides one data point for our ROC curve. This is a data point for the judge in a given experimental situation, when he has a given criterion level. If the experimental set-up is manipulated appropriately, the judge's criterion will change. His proportion of hits and false alarms will change. If a judge adopts a stricter criterion level then he will report fewer signals as being present. This will reduce both the proportion of hits and the proportion of false alarms. A further data point will then be available for the ROC curve. By manipulating the experimental conditions, a whole set of data points can be obtained, each one representing performance at a given criterion level. The curve drawn through these points is the ROC curve. It arches up symmetrically above the diagonal. The greater the signal strength the more the curve arches up. It is a routine matter to calculate d′ from the shape of the curve.

The important point to note is that a set of points, each corresponding to a given criterion level, is used to construct the curve. It is the shape of the curve that determines d′, not the exact position of the points. Another set of points corresponding to different criterion levels could be used and the same curve would result. Provided that there are enough points to get an accurate representation of the curve, its overall shape does not depend on which actual data points are used. Accordingly, the shape of the curve, and hence d′, does not depend on the actual criterion levels chosen. The value of d′ that is computed is thus independent of the actual criteria used to select the points on the curve; d′ is criterion-free. All that is required is that the criterion level is varied so that separate data points are obtained. A behavioral method can thus avoid criterion difficulties in two ways. First, it can set the criterion to an appropriate level by using a forced-choice procedure. Second, it can use judgments made at different criterion levels to compute an index of sensitivity.

The ROC curve takes a long time to construct, so d′ is not a measure that is obtained quickly. Each data point is the result of a separate experiment for a given criterion level. There are ways of speeding up the process. Instead of merely reporting whether he detects a signal or whether he feels there was just noise present, the judge could also add a sureness rating. He could say whether he was sure or unsure of his judgment. He could thus report "signal-sure", "signal-unsure", "noise-sure", or "noise-unsure". The judgment could be extended with categories such as "don't know but I will guess noise" and "don't know but I will guess signal". The number of categories can be extended as long as there remain an even number, so that the judge is forced to choose a signal or a noise. If there were a simple middle "don't know" category, less confident judges would be tempted to find refuge there. If forced to make a choice, a judge will often perform better than expected. Generally, four or six categories are used.

Using sureness ratings is rather like getting the judge to use several criteria simultaneously. To report a signal and be unsure is to report a signal using a less stringent criterion than if the judge reported he was sure of the signal. This signal detection rating procedure saves time because a whole curve, rather than a single data point, can be obtained during a single session. However obtaining the ROC curve can still be a lengthy process. With the rapid presentation possible for visual and auditory stimuli, the problem is not too extreme. For taste stimuli, the construction of an ROC curve can take days.[23]

2. P(A): The Proportion of Area Under an ROC Curve

Besides the time taken to calculate d', there are other difficulties. d' is a measure of the signal strength in terms of the standard deviation of the variation of the noise level, but when a signal is added, the noise level variation may change. The standard deviation of "signal + noise" might no longer be the same as that of noise alone. Having two standard deviations to deal with distorts the ROC curve and complicates the calculation of d'. It is not clear which standard deviation should be chosen as the unit of d'. Furthermore, an assumption was made that the variation of the noise level can be described by a normal distribution; this assumption is not always true. This means that the two assumptions upon which d' is founded do not always hold; d' is thus not always an applicable measure. For vision and hearing, the assumptions are often broken.[4] For the chemical senses, fewer data are available. For the two studies in taste that examined this issue,[23,24] the assumptions were broken in one of them.

Because of this, many researchers use an alternative measure of signal strength that does not depend on assumptions of normality or equality of standard deviation. This measure is derived from the shape of the ROC curve. As the signal strength increases, the ROC curve arches up further. Thus, more area is included below the curve. The proportion of area under the ROC curve is thus an alternative measure of signal strength. It is denoted by P(A).

3. Can We Now Forget About d' and P(A)?

Confronted by d' and P(A), it is not surprising that Sensory Evaluation has not embraced signal detection. An ROC curves takes a long time to construct and the measures obtained at the end are problematical. d' requires assumptions that do not always hold. P(A), the proportion of area under an ROC curve, is difficult to conceptualize. Middle management in a food company, without a background in communications engineering, is not going to take kindly to results couched in terms of areas under a curve that they cannot readily understand. All is not lost, however. More recent developments have produced a measure called the R-index. The R-index was developed independently for research into memory mechanisms[25] and turns out to be, by happy coincidence, the same measure as P(A). However, it is also a probability value and is simpler to conceptualize as such. It can therefore be used without any mention of ROC curves. In fact, all that needs to be remembered is that the R-index is the same as P(A) and because of this, it is called a signal detection measure. After that, signal detection and ROC curves can, in practice, be forgotten.

Before leaving these measures, it is worth getting a feel for them. Elliott[26] published tables of d' values corresponding to values of P(A). This means that after the ROC curve has been plotted, P(A) has been calculated and procedures have been used to ensure that the appropriate assumptions hold[4,22,23] a value of d' can be looked up in the tables.

When d' = 0, it means that the signal strength is no greater than the noise. If presented with both the signal and noise, a judge would have only a 50% probability of correctly picking the signal. This probability is equal to P(A). For d' = 1, this probability is raised to 76%. For d' = 1.81, the probability is 90%; for d' = 2.32, it is 95%, for d' = 2.9, it is 98%; while for d' = 3.28, it is 99%. Once the signal strength has a d' value of two or more, then, the detection rate will be relatively high.

For a given physical signal strength, judges with a higher d' or P(A) are more sensitive. For a given judge, a higher d' of P(A) means a stronger physical signal. Armed with these concepts, one can now examine an alternative approach to sensory difference testing.

IV. THE USE OF d' IN SENSORY DIFFERENCE TESTING

Frijters et al.[27] calculated values of d' for the triangle and 3-AFC testing procedures. Thus, from the proportion of correct responses, it is possible from his tables to find a d' value. Essentially, his table is an extension of Elliott's table of d' values from P(A).[26]

For a triangle test, if the proportion of correct responses is one third (33.33%), the judge is operating at chance levels, which means the signal strength (d') will be zero. Once the proportion reaches 60%, d' will be 1.98. For proportions of 70, 80, 90, and 99% and d' values will be 2.50, 3.13, 4.03, and 6.34, respectively.

The tables make a useful contribution to psychophysics and sensory evaluation, although it is important to remember that the use of d' rests on assumptions about the distribution of the noise and the signal + noise distributions. One should be sure of these assumptions before computing d'.

V. COMPUTATION OF THE R-INDEX

In our consideration of the R-index, we will first look at how to compute it. We will then discuss its statistical analysis and its application. First, the rating method will be applied to only two food treatments, then it will be modified for several food treatments. It is here that the advantages over scaling become apparent.

A. The R-Index by Rating: Two Food Treatments

First, the application of the R-index rating procedure to a simple difference test between two food treatments will be considered. This is the R-index in its simplest form and is most easily understood by considering the following example.

Let us assume that a judge is required to distinguish a flavor between two products, S and N. N could be a regular product while S could be a reformulated or differently processed version of this product. The aim of the test is to get a measure of the degree of difference between S and N without having to use a scaling procedure. In signal detection terminology, the task of the judge is to distinguish any flavor change signal in the food product S, from the background noise of the input from the food product N. We give the judge a given number of S samples and N samples in random order and require him to say whether each sample is S or N. He could make these judgments based on a session of practice at distinguishing between the two; he could make the judgments on the basis of "same as N", "different from N", with a standard N sample provided. He could even make the judgments in terms of "same as S", and "different from S". Should there be a predictable specific difference in flavor between S and N, this could be used as the basis for judgment. For example, should S have added sugar, the judge might be required to distinguish the samples in terms of "sweet" vs. "not sweet". There are many ways in which the judge may assign judgments to the S and N stimuli; the particular method will depend on the food and the testing circumstances.

The judge is thus required to distinguish which of the randomly presented samples are S and which are N. He is also required to say whether he is sure of his judgment or not. Often it may not be clear whether the product is, in fact, S or N; should the judge not be sure of his judgment, he is required to say so. Thus, each sample can be responded to as "definitely S" (S), "perhaps S but not sure" (S?), "definitely N" (N), or "perhaps N but not sure" (N?). This procedure, whereby the judge gives graded category responses, is called the rating procedure; it has been detailed previously in Section III.C.1. Let us assume that the judge is presented with ten N samples and ten S samples in random order. The exact number of samples is a matter of convenience for the specific task at hand; ten is chosen here merely to simplify the mathematics. Let us assume that the judge rates six of the S-samples as S, two as S?, and two as N?. Let us assume that he rates seven of the N samples as N, two as N?, and one as S?. His performance can be summarized by the following matrix.

		Judge's response				
		S	**S?**	**N?**	**N**	
Sample	S	6	2	2		$n_S = 10$
presented	N		1	2	7	$n_N = 10$

The question now becomes, "From this performance matrix, can the probability be estimated of the judge distinguishing between an S sample and an N sample?" The answer is yes; it can be done in the following manner:

Imagine that each S sample was presented in paired comparison with every possibly N sample; this would result in a total of 100 possible paired comparisons ($n_S \times n_T = 10 \times 10$). The question becomes one of predicting from the response matrix the number of these 100 possible theoretical paired comparisons in which the S and the N samples would be correctly identified. This percentage of correct paired comparisons is a good estimate of the probability of correctly distinguishing between S and N; 73% correct paired comparisons would mean that there is a 73% change ($p = 0.73$) of distinguishing between the two. This estimated probability value is the R-index. The chance level for correctly distinguishing between the two is 50%, while a probability of 100% would indicate perfect discrimination. Thus, the R-index ranges from 50% (chance) to 100% (perfect discrimination). The percentage of pair comparisons which would be correct can now be predicted.

Examining the response matrix, let us consider the six S samples identified as "definitely S" (S). When paired with any of the seven N samples identified as N, the two identified as N? and the one identified as S?, they would be correctly identified as S samples. Even the N samples rated as S? would not be chosen as S, because faced with a choice between a sample with a flavor rated S and one rated S?, the judge should sensibly choose the one rated S as the S sample. This gives $6 \times (1 + 2 + 7) = 60$ correct identifications, so far.

In the same way, the two S samples rates S? would be identified correctly in paired comparison with the 2 N samples rated N? or the seven N samples rated N. This gives a further $2 \times (2 + 7) = 18$ correct identifications. However, when these two S samples are compared to the N sample rated S?, the judge would not know which to choose as the S sample because they were both rated exactly the same (S?), so these two comparisons are scored as "don't know".

The two S samples rated N? would be identified correctly when compared to the seven N samples rated N, because given a choice whether a sample rated N? or N was the S sample, the one that was only doubtfully N (N?) would be the most likely to be S (score a further $2 \times 7 = 14$ correct). Comparison of these two N?-rated S samples with the two N samples rated N? would, on the other hand, leave the judge undecided (score another $2 \times 2 = 4$ don't knows). Thus, the predicted final tally of paired comparisons is 92 (60 + 18 + 14) correct identifications of S and 6 (2 + 4) "don't knows". Incidentally, the two S samples rated N? would be identified incorrectly (as N samples) when compared with the N samples rated S? (score 2×1 incorrect response). The total tally comes, as it should, to 100: 92 correct + 6 "don't knows" + 2 incorrect.

The judge, however, is not allowed a "don't know" response; he is forced to choose. It is assumed that when the judge is undecided, he must guess; half of these guesses on average will be correct and half incorrect. Thus, $6/2 = 3$ of the "don't knows" can be added to the correct scores giving a total of $92 + 3 = 95$. The calculation is outlined below:

Judge's response

		S	S?	N?	N	
Sample	S	6	2	2		$n_S = 10$
presented	N		1	2	7	$n_N = 10$

Number correct:	$6 \times (1 + 2 + 7)$	$= 60$
	$2 \times (2 + 7)$	$= 18$
	2×7	$= 14$
Don't know:	2×1	$= 2$
	2×2	$= 4$

$$R = 60 + 18 + 14 + \tfrac{1}{2}(2 + 4) = 95$$

Thus, from the response matrix the prediction would be that the judge would correctly distinguish between S and N on 95 out of 100 paired comparisons; the estimated probability of distinguishing S and N is thus 95%, or 0.95. This estimated probability is called the R-index and, as such, is a useful measure of the degree of difference between S and N for this judge.

In the example there were ten replications given for both S and N samples; any convenient number of samples may be chosen, bearing in mind that the larger the number chosen, the more representative the sample will be of the judge's behavior.

In general, the computational steps just explained can be summarized by Equation 1:

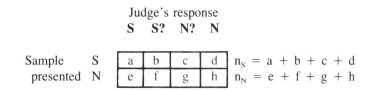

The R-index is given by:

$$R = \frac{a(f + g + h) + b(g + h) + ch + 1/2(ae + bf + cg + dh)}{n_S\, n_N} \tag{1}$$

In the example above, we used four response categories (S, S?, N?, N). More could be used to obtain greater resolution. Addition of "don't know but will guess S" (S??) and "don't know but will guess N" (N??) is possible without undue complication of the judge's task.[23,24] Addition of categories is possible, as discussed in Section III.C.1., as long as a central "don't know" category is avoided. With such a response category, less confident judges would be tempted to keep responding "don't know" when they could in fact distinguish S and N. Thus, with the rating procedure there should always be an even number of categories. Most of the information is obtained using the four categories in the example. Six categories will provide a little more information but generally four are satisfactory. No less than four categories should be used, however, otherwise too little information will be obtained for a good probability estimate.

When more categories are used, the calculation of the R-index is extended using the same strategy as before. Basically, every number in the top (signal) row of the matrix is multiplied by all of the numbers to the right in the bottom (noise) row, as with a f + g + h + . . . etc.; with more categories there will simply be more numbers in the bottom row. The total of these values is taken. Then, every number is multiplied by the number directly below it, and half the total of these values is taken, as with $^1/_2$(ae + bf + . . . etc.). These two totals are added together and the sum divided by $n_S n_N$, the product of the total of the numbers in the top (signal) row and the bottom (noise) row. This gives the R-index as a fractional value; multiply by 100 to get the percentage.

The R-index rating procedure provides a numerical measure of degree of difference between two foods S and N in terms of a probability of correctly picking S when it is presented in paired comparison with N. The numerical measure is obtained without the judge having to use any scaling procedure; he merely has to decide whether he is sure or not of his judgment. Very often, such a degree of precision will not be necessary. When two food treatments are compared, a dichotomous "same" vs. "different" result at a given level of significance will often suffice. This is readily obtained by the regular paired comparison, triangle, or duo-trio tests.

Some advantage can be gained, however, should a further comparison be made between the product N and a further product, say, S_1. In this case, numerical indexes would be available to compare the difference between N and S and between N and S_1. It is here that the advantage of the R-index begins to be apparent. Should several products S_1, S_2, and S_3 be compared with product N, R-index values would be convenient measures of the degree of difference. In practice, the stimuli would not all be compared in separate experiments; a shorter approach, outlined in Section V.C, would be used.

One objection to the R-index procedure might be its length. Although it does not take a great deal of time to taste sufficient samples to calculate an R-value, it certainly takes longer than two to three triangle tests. For Sensory Evaluation I, the R-index is perfectly appropriate. Here, a few sensitive judges are used to sample as much of the food treatments as possible. Considerable time should be allotted for each judge. For Sensory Evaluation II, an R-index measure for each individual judge might take too long, although this depends on the nature of the judgment.

The R-index is a signal detection measure. It is a matter of simple geometry to show that Equation 1 is equivalent to an expression for the area under an ROC curve. Thus, for a given signal-noise pair, the R-index is equivalent to P(A). In fact, such a probability estimate was discussed by Green and Swets[4] and given the symbol P(C).

B. The R-Index as a Threshold Measure

Because threshold measures so often involve the comparison of a stimulus with a blank and can be considered to be difference tests (see Section II.D.), the R-index rating procedure can also be applied here. If S is a given concentration of stimulus and N the "blank", the same procedure can be used to determine the judge's sensitivity to that stimulus concentration. This is a measure of his sensory sensitivity.

A threshold is a concentration at which a stimulus can be distinguished from a blank at some specified level of performance, say, correctly sorting four stimuli from four blanks.[9] The threshold will be expressed in terms of the minimum concentration at which this specified level of performance is possible. The R-index measure of sensitivity takes a different approach. For a given concentration, it measures the degree to which the stimulus can be distinguished from the blank (R-index). For example, the R-index for a given concentration of one stimulus may be higher than for the same concentration of a second stimulus; the judge is thus less sensitive to the second stimulus.

The R-index measure of sensitivity has the advantage of simplicity and rapidity. For example, with NaCl taste sensitivity, it takes only a moment or two to taste ten water samples and ten 3-mM NaCl samples in random order and respond with the judgment "salt" vs. "water", "sure" vs. "not sure". It can take an hour to measure an NaCl taste threshold by the standard methods.

In the past few years, the matter of taste sensitivity has been largely neglected in favor of measures of the intensity of suprathreshold stimuli. One reason for this may have been the relative convenience of intensity measures compared with the lengthy and involved threshold measures. However, with the rapid R-index procedures available, issues necessitating rapid measurement of sensitivity per se can now be addressed directly.[6,28,29] The procedure can also be adapted to psychophysical procedures other than threshold-type judgments involving the absence or the presence of a stimulus. Measures of "mixedness" vs. "singularity" of a series of single stimuli and mixtures have also been made using R-index procedures; these were used as part of a study of the degree of blending of taste mixtures.[30]

C. The R-Index by Rating: More Than Two Food Treatments

As mentioned before, the R-index would seem most useful in sensory difference testing when the degree of difference is to be measured among several food products. Following our example, we might be interested in several reformulations of a product S_1, S_2, and S_3 and their degree of difference from the original product N. Such information could then be used to guide further reformulation; it could indicate whether style S_1, S_2, or S_3 were preferable approaches to reformulation. Should any of the reformulations be indistinguishable from N, it might even be decided to send the product to market. Traditionally, the approach would be first to use difference tests to establish whether S_1, S_2, or S_3 differed from N. A scaling procedure would then be used to determine the degree of difference between the products, with the resulting complications discussed earlier (see Section II.E).

From our discussion of the R-index so far, an alternative approach would be to determine the degree of difference between N and each reformulation using the R-index rating procedure. This would necessitate a separate experiment to compare S_1 with N, S_2 with N, and S_3 with N. However, the procedure can be streamlined. S_1, S_2, S_3, and N could all be tested in random order in one session and the judge required to state whether each is S (a reformulation) or N (the original product). Again, the actual details of the experiment can vary. The judge may be so familiar with N that he could recognize any variation from the true flavor and deem it an S, along with a "sure" or "not sure" judgment. On the other hand, he may prefer to taste a standard N before making each judgment; in this case his judgment boils down to saying that it is the "same as" or "different from" the standard. An approach could be used whereby a warm-up procedure[23] was used to teach the judge the differences between S and N. All variants are possible; they do not affect the computation of the R-index. All that is needed is a series of judgments categorizing the stimuli as S or N with an additional "sure" or "not sure" statement. How these are obtained and how many can be derived from a single session will surely vary from product to product.

Consider the computation; assume that the judge tastes in random order ten samples of a reformulation of a product: S_1, S_2, and S_3, as well as ten samples of the original N. Again, ten is chosen only to make the mathematics simple; the actual number sampled depends on the degree of resolution desired and the amount of time and product available. The judge categorizes the samples as before and we can represent the results on a response matrix. This time, however, the matrix would have not two rows (S and N) but four rows (S_1, S_2, S_3, and N); such a matrix is given below.

Judge's response

	S	S?	N?	N
Sample presented S_3	9	1		
S_2	6	3	1	
S_1	6	2	2	
N		1	2	7

S_3/N, R = 99.5%
S_2/N, R = 96.5%
S_1/N, R = 95.0%

R-index values can be calculated for the probability of distinguishing S_1 from N, S_2 from N, and S_3 from N in the same way as before (in fact, comparisons could also be made between any pairs of S_1, S_2, and S_3, if required). For S_1 (6,2,2) and N (1,2,7) the R-index is 95% (these are the same values as in the two-sample example). For S_2 and N, R = 96.5%, and for S_3, and N, R = 99.5%.

Thus, in this example, the judge could distinguish between the reformulations and the regular product nearly all the time; the reformulations were fairly distinct from the original product. S_3 was the most different from N, S_1 the least different.

How to determine whether any of these R-indexes are above chance will be discussed later. The values here are obviously above the chance level of 50%, but other cases may not be so plain. It generally requires statistical procedure to determine whether an R-index of 55% is significantly greater than the chance 50% level.

An interesting variant that should be built into experimental designs, if possible, is to make an experimental determination of the chance level. When the sample of tastings is small (say only ten of each product), it is possible that a judge who is detecting no difference and thus working at chance levels may get an R-index of approximately, say, 60% simply by chance. For so few tastings of a product, the variance of the R-index can be large and "high chance" values are possible. One way to estimate how high chance values can become in a given experiment, is to arrange that the product S_1, instead of being a reformulation, is exactly the same as N. Theoretically, if enough tastings were made, the R-index between S_1 and N would then be 50%. If it rose as high as 60%, it would indicate that only R-indices greater than this value indicate that the judge could differentiate the two products better than chance. It is a nonstatistical way of determining chance performance.

In psychophysical measures, reversals can occur with less experienced judges. When a judge is tasting, say, distilled water and 3 mM NaCl, he may confuse the two. NaCl can be tasteless, and an inexperienced judge may think that tastelessness signals the presence of water. Distilled water can have a definite taste and the inexperienced judge may think that this signals the presence of salt.[6] In this case, the judge would be judging water as more like salt than 3 mM NaCl. If water were the noise and NaCl the signal in the R-index calculation, a value below 50% would occur. This difficulty can be overcome by allowing the judge greater familiarity with the stimuli.

The exact significance of an R-index below 50% is usually easy to determine. It may fall slightly below 50% by chance, it may mean that the judge has confused the stimuli, or it can mean that the signal has less of a given characteristic than the noise. What an R-index below 50% indicates generally becomes obvious after examining the data, the experimental design, and the general trends among judges.

D. The R-Index is not a Substitute for Intensity Scaling

It is important to realize that the R-index procedures are designed for difference testing. They are applicable in situations where products are sufficiently similar to warrant difference tests. In such measurements, judges would make errors and R-indexes would fall below 100%.

Should the R-index procedure be mistakenly used in a situation where all the products were clearly different, all R-indices would register this by giving values of 100%. The degree of difference would not be indicated. Clearly a scaling procedure would have been more appropriate.

It is important with any behavioral procedure to be cautious in its application and not see it as a panacea for all sensory analysis. The R-index is merely an addition to the sensory testing procedures already available.

VI. STATISTICAL TREATMENT OF R-INDEX VALUES

A. Analysis of Variance

The R-index is a numerical measure of the degree of difference and is thus susceptible to parametric statistical analysis. The statistical analysis used, however, will depend on the aims of the experiment.

In an experiment where a group of judges make judgments about a set of products (for example, the degree of difference between S_1 and N, S_2 and N, and S_3, and N), each judge will supply an R-index value for each product (one each for S_1, S_2, and S_3). A mean value of the R-indexes from all of the judges can be calculated for each product. To determine whether these mean R-indexes are significantly different, two-factor (products and judges) analysis of variance with the appropriate multiple comparisons would be appropriate.[16] Instead of products, the treatments may be different experimental conditions. The experiment may be comparing taste sensitivity under different conditions as part of a program of psychophysical research. The same statistical treatment would be appropriate.

B. Rank Sums Test

For Sensory Evaluation I (Section II.B.1), each judge would be examined separately. The question becomes a matter of whether a specific R-index indicates that the judge can differentiate the products better than chance. It is a matter of inspecting a judge's matrix and determining whether it indicates significant differentiation or not. The appropriate statistical test is the Rank Sums Test.[16,32] The test apparently was devised by Frank Wilcoxon and has been called the Rank Sums Test or the Wilcoxon Rank Sums Test. It is an independent samples test and should not be confused with the related samples Wilcoxon Test.[16] Also, because it is equivalent to Mann and Whitney's U-Test, it is sometimes called the Mann-Whitney Test. Faced with this plethora of names, it will be called the Rank Sums Test in this chapter.

The Rank Sums Test is simply a matter of computing a value from the numbers in the R-index matrix. Using this value, tables are consulted to determine whether the R-index indicates significant differentiation between the two products or not. The Rank Sums test has its own set of tables,[16,32] but happily, when it is applied to the R-index matrix, the circumstances demand that significance be tested using the common normal distribution tables. The probability obtained with these tables gives the probability of getting an R-index this large on the null hypothesis. The normal distribution tables are often used with nonparametric tests such as the Rank Sums Test. Under particular circumstances, the distributions of these tests approximate to normal.

More specifically, the statistical procedure involves calculation of a z-value. Normal distribution tables are used to provide a probability value associated with the z-value. This is the probability of getting an R-index value this large on the null hypothesis. The z-value is given by the following equation:[16,32]

$$z = \frac{S - K}{SD} \tag{2}$$

where S is the Rank Sums Index, K is a correction for continuity, and SD is a standard deviation value.

A general formula will first be given for computing the z-value, then the procedure will be illustrated with the example given in Section V.A. The values in the R-index matrix can be represented in a general way, as follows:

	S	S?	N?	N	
S	a	b	c	d	$n_S = a + b + c + d$
N	e	f	g	h	$n_N = e + f + g + h$

Firstly, S is calculated from Equation 3:

$$S = |u - u'| \tag{3}$$

where u is the total of the products of each score in the top (S) row times each score to the right of it in the botton (N) row and u' is the total of the products of each score in the top (S) row times each score to the left of it in the bottom (N) row.
Here

$$u = a(f + g + h) + b(g + h) + ch \tag{4}$$

$$u' = d(e + f + g) + c(e + f) + be \tag{5}$$

Second, SD is calculated from the column totals (T) and the row totals (n). The column totals from left to right are

$$T_S = a + e$$
$$T_{S?} = b + f$$
$$T_{N?} = c + g$$
$$T_N = d + h$$

$$SD = \sqrt{\frac{n_s n_N[(n_s + n_N)^3 - \Sigma\, T^3}{3(n_s + n_N)(n_s + n_N - 1)}} \tag{6}$$

Third, K is calculated from the left hand column total (T_S), the right hand column total (T_N), and C (the number of categories in the matrix; here, C = 4).

$$K = \frac{2(n_s + n_N) - T_s - T_N}{2(C - 1)} \tag{7}$$

Finally, a z-value is calculated from S, K, and SD, using Equation 2, then normal probability tables[16,33] are used. The probability associated with a z-value this large or greater can be found from the tables. This probability will equal the one-tailed probability of getting an R-index this large by chance. This value is doubled to get the two-tailed probability. This statistical analysis will now be illustrated with the example from Section V.A. The R-index matrix is given below:

	S	**S?**	**N?**	**N**	
S	6	2	2	0	$n_S = 10$
N	0	1	2	7	$n_N = 10$

R-index = 95%

First, S is calculated from Equations 3, 4, and 5.

$$u = 6(1 + 2 + 7) + 2(2 + 7) + 2 \times 7$$
$$= 60 + 18 + 14 = 92$$
$$u' = 0(0 + 1 + 2) + 2(0 + 1) + 2 \times 0$$
$$= 2$$
$$S = |u - u'| = 92 - 2 = 90$$

A negative value for u − u′ merely indicates that the R-index is a value below 50%. For this calculation the modulus (positive value) is always taken.

SD is then calculated. First, the column totals are noted.

$$T_S = 6 + 0 = 6 \qquad T_S^3 = 216$$

$$T_{S?} = 2 + 1 = 3 \qquad T_{S?}^3 = 27$$

$$T_{N?} = 2 + 2 = 4 \qquad T_{N?}^3 = 64$$

$$T_N = 0 + 7 = 7 \qquad T_N^3 = 343$$

$$\text{Total } \Sigma T^3 = 650$$

The values are then substituted in Equation 6.

$$SD = \sqrt{\frac{n_s n_N [(n_s + n_N)^3 - \Sigma T^3]}{3(n_S + n_N)(n_S + n_N - 1)}}$$

$$= \sqrt{\frac{100[20^3 - 650]}{3 \times 20 \times 19}}$$

$$= \sqrt{\frac{100 \times 7350}{1140}} = 25.39$$

K is finally calculated from Equation 7.

$$K = \frac{2(n_S + n_N) - T_S - T_N}{2(C - 1)}$$

$$= \frac{2 \times 20 - 6 - 7}{2(4 - 1)} = \frac{27}{6} = 4.5$$

Finally, a z-value is calculated using Equation 2.

$$z = \frac{S - K}{SD}$$

$$= \frac{90 - 4.5}{25.39} = 3.37$$

The normal distribution probability associated with a z-value of 3.37 can be obtained from tables readily available in most texts. The probability of obtaining a z-value of 3.37 or larger is 0.0004. This is a one-tailed probability; it is doubled to obtain the two-tailed probability, $2 \times 0.0004 = 0.0008$. Thus, the probability of obtaining an R-index value as large as 95% for this number of tastings ($n_S = n_N = 10$) on the null hypothesis is 0.0008. The null hypothesis is thus rejected, so the judge has demonstrated significant evidence of differentiating between the products S and N.

This calculation is long and detailed and is best relegated to a computer. The Appendix gives a computer program in BASIC that will give a value of the R-index and derive from the Rank Sums Test and the Normal distribution the one-tailed and two-tailed probabilities of obtaining an R-index this large on the null hypothesis.

C. Statistical Analysis and the Design of Sensory Analytic Tests

For Sensory Evaluation I, the R-index values for each judge will be analyzed separately. The appropriate test is the Rank Sums Test.

For Sensory Evaluation II, the situation is more complicated. It is important, here, that the judge remains typical of the consumer regarding his ability to differentiate between the food products. For the judges to learn to identify the products for the R-index rating procedure, the training and practice needed could give them familiarity with the product that was greater than that of ordinary consumers. This would defeat the purpose of the test. The use of a standard might help prevent this. The judges could then categorize the stimuli as "same as" or "different from" the standard. This would help prevent the overfamiliarization involved when the judges learn to identify the products from memory, yet it could be argued that the replicate tastings required for each judge to produce an R-index might provide him with practice in skills atypical of a consumer, even when steps are taken to avoid excessive stimulus learning, although this may be said about any difference testing procedure involving replicate testings. The question of overfamiliarzation must really be decided for each food and for each type of judgment. Experimenters should use their experience with the product and their knowledge of the consumers to decide whether the testing procedure gives the judges too much familiarity with the product.

Generally, however, for Sensory Evaluation II the traditional difference tests would seem to suffice. The aim of the measurement is to give brief tests to prevent greater than usual familiarization with the stimuli. They should then be given to as many judges as possible, because it is the judges who are being sampled. However, should numerical measures of degree of difference be necessary, a decision must be taken. The experimenter must decide whether to calculate an R-index for each judge. The mean R-index values, giving the degree of difference found by the judges in the sample, would then give a good estimate of the mean values for the population, giving a measure of the degree of differentiation in the population. Analysis of variance with multiple comparisons would indicate which R-indices would differ significantly in the population. However, the experimenter may still be reluctant to give so many judgments to each panelist. One way around the problem would be to calculate a composite R-index for a given panel. Each judge could be given very few tastings (say, one or two ratings for each food treatment). The data for each judge could then be pooled in a composite matrix for the whole panel, care being taken to ensure that each judge contributed an equal amount of data. A composite panel R-index could then be calculated. The difficulty here is that different judges have different criteria so that the lines between "sure" and "not sure" would be somewhat blurred. This may tend to give more ties than obtained from a single judge thus lowering the R-index. This composite R-index could then be tested for significance with the Rank Sums Test. It could be argued logically that if a representative panel differentiated foods to a certain degree, the population would do so also.

VII. APPLICATION TO CONSUMER PREFERENCE TESTING

As with Sensory Evaluation II, obtaining preference data from consumers is a task that must be done without giving the consumer undue practice, which would render him atypical. Much consumer testing requires the use of complex hedonic scaling techniques. In the hands of an unpracticed consumer, the data from such scales would be expected to be approximate. A complex parametric statistical analysis on such data may look elegant but convey little. With consumer testing it would seem sensible to use simple behavioral measures to match the consumer's lack of psychophysical skill and brief tests to allow a wide sampling of judges. Paired preference or ranking for preference would seem a favorable method.

Should some form of rating be used, a composite panel matrix could be constructed to calculate panel R-index values, following the approach suggested for Sensory Evaluation II. Such R-index values would express the degree of preference among a set of products relative to some standard product.

VIII. SUMMARY

The R-index, its computation, and its statistical analysis have been discussed with special reference to sensory evaluation. The use of the R-index and its statistical treatment varies according to whether the task at hand is Sensory Evaluation I or II. The technique has been used to solve the problem of getting numerical measures of the degree of difference between food treatments[34-36] and has considerable psychophysical use. However, like all new measurements, it is in danger of being used in inappropriate circumstances or of being viewed as a panacea. It is merely a useful technique which solves the problem of generating numerical estimates of degree of difference. It allows new possibilities for sensory difference testing. It is a sufficiently flexible technique to allow the data to be collected with a wide variety of behavioral measures and food presentation procedures so as to fit the particular needs of particular products. It provides an approach to sensory analysis which will yield the best results after the experimenter has modified the techniques to suit his or her own needs. It has been used successfully in several countries for diverse products ranging from whisky to colas, from popsicles to dairy products, and from fresh fruit to cigarettes.

X. APPENDIX

COMPUTER PROGRAM FOR COMPUTATION AND STATISTICAL ANALYSIS OF THE R-INDEX

Magdalena Tamura*

This program calculates an R-index from matrix values and tests its significance using the Rank Sums Test. It first requests the number of categories in the matrix (generally four: S, S?, N?, and N). The numbers for the top row of the matrix are entered and then the numbers for the bottom row. The completed matrix is shown on the screen and the opportunity given to correct any errors. If there are errors, the program will ask whether each row is correct and invite correction. Having completed the matrix correctly, the R-index value is computed along with the Rank Sums Test z-value and the probabilities of obtaining R-index values this large on the null hypothesis (one- and two-tailed). The program then provides the choice of further R-index computations with the same or different numbers of categories or termination of the program.

The program is detailed first in BASIC, then in FORTRAN. Finally, a sample run is given.

* Department of Food Science and Technology, University of California, Davis.

BASIC

```
C===========================================================
C                      R-INDEX
C             MAGDALENA TAMURA   [8-AUG-85]
C===========================================================
C
        DIMENSION R1(10),R2(10)
C
        WRITE (5,55) '********************************'
        WRITE (5,55) 'WELCOME TO THE R-INDEX CALCULATION'
        WRITE (5,55) '********************************'
55      FORMAT (10X,A34/)
100     WRITE (5,700)
700     FORMAT(1X,'ENTER THE NUMBER OF CATEGORIES IN THE R-INDEX MATRIX:')
        READ(5,*) K
        WRITE (5,4)
4       FORMAT (/)
33      TYPE*,'ENTER THE NUMBERS IN THE TOP ROW OF THE MATRIX FROM'
        TYPE*,'LEFT TO RIGHT, SEPARATED BY COMMAS. Include zeros.'
        TYPE*,'(No comma is required after the last number)'
        READ (5,*) (R1(I),I=1,K)
        WRITE (5,4)
        TYPE*,'ENTER THE NUMBERS IN THE BOTTTOM ROW:'
        READ (5,*) (R2(I),I=1,K)
        WRITE (5,4)
        WRITE(5,39)'THE R-INDEX MATRIX WITH',K, 'CATEGORIES IS:'
39      FORMAT (1X,A23,2X,I1,2X,A14/)
81      TYPE*,'----------------------------------------------'
        WRITE (5,14)(R1(I),I=1,K)
        WRITE (5,14)(R2(I),I=1,K)
14      FORMAT (1X,10(F4.0,2X))
        TYPE*,'----------------------------------------------'
        WRITE (5,4)
        TYPE*, 'IF THIS IS CORRECT TYPE        Y'
        TYPE*, 'IF THIS IS NOT CORRECT TYPE    N'
        READ (5,9)ANS
9       FORMAT (A1)
        IF((ANS.EQ.'N').OR.(ANS.EQ.'n')) GOTO 850
        IF((ANS.EQ.'Y').OR.(ANS.EQ.'y')) GOTO 750
850     WRITE (5,4)
        TYPE *,'IS THE TOP ROW CORRECT? (Y/N)'
        READ(5,9) ANS
        IF ((ANS.EQ.'N').OR.(ANS.EQ.'n')) GOTO 67
        IF ((ANS.EQ.'Y').OR.(ANS.EQ.'y')) GOTO 60
67      WRITE (5,4)
        TYPE*,'RE-ENTER THE TOP ROW FROM LEFT TO RIGHT'
        READ (5,*) (R1(I),I=1,K)
60      WRITE (5,4)
        TYPE*,'IS THE BOTTOM ROW CORRECT? (Y/N)'
        READ (5,9) ANS
        IF ((ANS.EQ.'N').OR.(ANS.EQ.'n')) GOTO 77
        IF ((ANS.EQ.'Y').OR.(ANS.EQ.'y')) GOTO 79
77      WRITE (5,4)
        TYPE*, 'RE-ENTER THE BOTTOM ROW FROM LEFT TO RIGHT'
        READ (5,*) (R2(I),I=1,K)
79      WRITE (5,4)
        WRITE (5,80)'THE CORRECTED R-INDEX MATRIX IS:'
80      FORMAT (1X,A32,/)
        GOTO 81
C
C       R-INDEX CALCULATION
C       -------------------
750     XTOTL = 0.
        DO 140 I=1,K-1
                SUM=0.
                DO 20 J=I+1,K
20                      SUM=SUM+R2(J)
140             XTOTL=XTOTL+R1(I)*SUM
        ASUM = 0.
        DO 300 I=1,K
300             ASUM=ASUM+R1(I)*R2(I)
        RN1 = 0
        RN2 = 0
```

```
C         Z-VALUE CALCULATION
C         -------------------
          YTOTL =0.
          DO 500 I=K,2,-1
                XSUM =0.
                DO 99 J=1,I-1
99                    XSUM=XSUM+R2(J)
500           YTOTL=YTOTL+R1(I)*XSUM
          S = ABS(XTOTL-YTOTL)
          RN = RN1+RN2
          C=(2.*RN-(R1(1)+R2(1)) - (R1(K)+R2(K)))/(2.*(K-1))
          CSUM=0.
          DO 900 I=1,K
900           CSUM=CSUM+(R1(I)+R2(I))**3
          SD = SQRT(RN1*RN2*(RN**3-CSUM)/(3.*RN*(RN-1)))
          Z= (S-C)/SD
C
          WRITE (5,4)
          TYPE *,'----------------------------------'
          WRITE (5,15)'THE R INDEX IS:',R,' OR ', R*100
15        FORMAT (1X,A15,2X,F6.4,A4,F6.2)
          TYPE *,'----------------------------------'
          WRITE (5,4)
          WRITE (5,16)'THE RANK SUMS Z VALUE IS:',Z
16        FORMAT (1X,A25,2X,F6.4,/)
          IF ((Z.GT.4.0).OR.(Z.LT.-4.0)) GOTO 956
          CALL NORMAL(Z,PROB)
          TYPE*,'THE PROBAB. OF GETTING AN R THIS LARGE '
          TYPE*, 'ON THE NULL HYPOTHESIS IS:'
          WRITE (5,51)'ONE-TAILED =',PROB
51        FORMAT (4X,A13,1X,F7.5)
          WRITE (5,51)'TWO-TAILED =',PROB*2
          WRITE (5,4)
          GOTO 65
956       TYPE*,'THE PROBAB. OF GETTING AN R THIS LARGE '
          TYPE*,'ON THE NULL HYPOTHESIS IS:'
          WRITE (5,515)'ONE-TAILED ( .00003'
          WRITE (5,515)'TWO-TAILED ( .00006'
515       FORMAT (4X,A19)
          WRITE (5,4)
65        WRITE(5,301)'ENTER 1 TO CONTINUE R-INDEX CALCULATIONS'
          WRITE (5,302)'WITH',K,'CATEGORIES?'
301       FORMAT (1X,A40)
302       FORMAT (9X,A4,I2,1X,A11)
          TYPE*,'ENTER 2 TO CONTINUE WITH DIFFERENT NUMBER OF CATEGORIES'
          TYPE*,'ENTER 3 TO EXIT PROGRAM'
          READ (5,9)ANS
          IF (ANS.EQ.'1')GOTO 33
          IF (ANS.EQ.'2') GOTO 100
          IF (ANS.EQ.'3') STOP
          END
```

FORTRAN

```
C=================================================
C       SUBROUTINE FOR FORTRAN R-INDEX PROGRAM
C       SIMPSON'S 1/3 RULE INTEGRATION
C               NORMAL CURVE
C=================================================
        SUBROUTINE NORMAL(Z1,PROB)
        DATA N  /2/
        DATA ODD,EVEN    /0.,0./
        DATA Z2,POLD     /4.0,0./

        SUM = F(Z1)+F(Z2)
100     DELT = ABS (Z2-Z1)/FLOAT(N)
        Z = Z1 +DELT
        ODD = F(Z)
        DO 25 J=1,N-1,2
                Z = Z+2*DELT
25              ODD = F(Z) + ODD
        PROB = DELT/3.*(SUM+4.*ODD+2.*EVEN)
        IF (ABS(PROB-POLD).LE..001)GOTO 50
        EVEN= EVEN +ODD
        POLD = PROB
        N=N*2
        GOTO 100
50      PROB=PROB+.00003
        RETURN
        END


C
        FUNCTION F(X)
C
        F = EXP(-0.5*X**2)/SQRT(2.*3.141592)
        RETURN
        END
```

SAMPLE RUN

```
10 REM ************************************************************
20 REM *****************          R - INDEX          ***************
30 REM ************************************************************
40 REM **************          MAGDALENA TAMURA          ***************
50 REM **************          12-AUG-85          ***************
60 REM ************************************************************
70 REM
80 REM ~~~   FNF(X) == NORMAL DISTRIBUTION EQUATION   ~~~~~~~
90 DEF FNF(X)=EXP (-.5*X^2)/SQR(2*3.141592)
100 REM
110 CLS
120 PRINT"************************************************************"
130 PRINT"**************  WELCOME TO THE R-INDEX CALCULATION  ***********"
140 PRINT"************************************************************"
150 PRINT: PRINT
160 PRINT:INPUT "Enter the number of categories in the R-index matrix: ";K
170 DIM R1(K),R2(K)
180 PRINT:PRINT"Enter the scores in the top row of the matrix (include zeros)."
190 FLAG=0
200 PRINT
210 FOR I = 1 TO K
220     PRINT "SCORE";I;": ";
230     INPUT"",R1(I)
240 NEXT I
250 IF FLAG=1 GOTO 470
260 PRINT
270 PRINT: PRINT "Enter the scores in the bottom row of the matrix."
280 PRINT
290 FOR I = 1 TO K
300     PRINT "SCORE";I;": ";
310     INPUT"",R2(I)
320 NEXT I
330 PRINT
340 PRINT "---------------------------------------------------------------"
350 PRINT "TOP ROW (S):";
360 FOR I = 1 TO K-1:PRINT USING "###   ";R1(I);
370 NEXT I
380 PRINT USING"###   "; R1(K)
390 PRINT "BOT ROW (N):";
400 FOR I = 1 TO K:PRINT USING"###   ";R2(I);
410 NEXT I
420 PRINT "---------------------------------------------------------------"
430 PRINT:INPUT"Are all these data values correct? (YES OR NO): ";O$
440 IF O$="YES" OR  O$="yes" OR O$="y" OR O$="Y" THEN GOTO 500
450 PRINT: INPUT "Is the top row correct? (YES OR NO): ";O$
460 IF O$="NO" OR  O$="no" OR O$="N" OR O$="n" THEN FLAG=1: GOTO 200
470 PRINT: INPUT "Is the bottom row correct? (YES or NO): ";O$
480 IF O$="NO" OR  O$="no" OR O$="N" OR O$="n" THEN GOTO 260
490 REM
500 REM ~~~~~~~~~~~~~  R-INDEX CALCULATIONS ~~~~~~~~~~~~~~~
510 REM ~~~~~~~~~~~~~~~~~~~~~~~~~~~~~~~~~~~~~~~~~~~~~~~~~~
520 XTOTL = 0
530 FOR I = 1 TO K-1
540     SUM = 0
550     FOR J = I+1 TO K
560         SUM = SUM + R2(J)
570     NEXT J
580     XTOTL = XTOTL +R1(I) * SUM
590 NEXT I
600 ASUM = 0
610 FOR I = 1 TO K
620     ASUM=ASUM + R1(I)*R2(I)
630 NEXT I
640 RN1 = 0
650 RN2 = 0
660 FOR I = 1 TO K
670     RN1 = RN1 + R1(I)
680     RN2 = RN2 + R2(I)
690 NEXT I
700 R = (XTOTL +ASUM * .5) / (RN1*RN2)
710 REM
```

```
720 REM ~~~~~~~~~~~   Z-VALUE CALCULATION   ~~~~~~~~~~~
730 REM ~~~~~~~~~~~~~~~~~~~~~~~~~~~~~~~~~~~~~~~~~~~~~~~~
740 YTOTL = 0
750 FOR I = K TO 2 STEP -1
760    XSUM = 0
770    FOR J = 1 TO I-1
780       XSUM = XSUM + R2(J)
790    NEXT J
800    YTOTL = YTOTL+R1(I) *XSUM
810 NEXT I
820 S = ABS(XTOTL - YTOTL)
830 RN = RN1 +RN2
840 C = (2 * RN - (R1(1)+R2(2)) - (R1(K)+R2(K)))/ (2*(K-1))
850 CSUM = 0
860 FOR I = 1 TO K
870    CSUM = CSUM +(R1(I) + R2(I)) ^3
880 NEXT I
890 SD = SQR(RN1*RN2*(RN^3-CSUM)/(3*RN*(RN-1)))
900 Z1 = (S-C)/SD
910 REM
920 REM ~~~~~~~~~~          OUTPUT RESULTS      ~~~~~~~~~~~~~~~~~
930 REM ~~~~~~~~~~~~~~~~~~~~~~~~~~~~~~~~~~~~~~~~~~~~~~~~~~~~~
940 PRINT:PRINT"--------------------------------"
950 PRINT"THE R-INDEX IS: ";:PRINT USING "*.****";R;:PRINT "  OR  ";
960 PRINT USING "**.**";R*100;:PRINT " %"
970 PRINT"--------------------------------"
980 PRINT ""
990 PRINT:PRINT"THE RANK SUMS TEST Z VALUE IS: ";Z1
1000 IF Z1>4 OR Z1<-4 GOTO 1090
1010 GOSUB 1240
1020 PRINT:PRINT "THE PROBABILITY OF GETTING AN R THIS LARGE"
1030 PRINT"ON THE NULL HYPOTHESIS IS:"
1040 PRINT "     ONE-TAILED = ";
1050 PRINT USING "*.*****";PROB
1060 PRINT "     TWO-TAILED = ";
1070 PRINT USING "*.*****";PROB*2
1080 GOTO 1140
1090 PRINT:PRINT "THE PROBABILITY OF GETTING AN R THIS LARGE"
1100 PRINT"ON THE NULL HYPOTHESIS IS:"
1110 PRINT "     ONE-TAILED < .00003 "
1120 PRINT "     TWO-TAILED < .00006 "
1130 REM ~~~~~~~    MENU   ~~~~~~~~~
1140 PRINT:PRINT:PRINT " ENTER 1 TO CONTINUE R-INDEX WITH ";K;" CATEGORIES"
1150 PRINT" ENTER 2 TO CONTINUE R-INDEX WITH DIFFERENT NUMBER OF CATEGORIES"
1160 PRINT" ENTER 3 TO EXIT PROGRAM"
1170 PRINT: INPUT "    OPTION NUMBER: ",CHOICE
1180 ON CHOICE GOTO 180,1190,1420
1190 CLS:ERASE R1,R2:GOTO 160
1200 REM
1210 REM ~~~~~~~~~~   AREA UNDER NORMAL DISTRIBUTION        ~~~~~~~~~~~
1220 REM ~~~~~~~~~   SIMPSON'S 1/3 RULE OF INTEGRATION      ~~~~~~~~~~~
1230 REM ~~~~~~~~~~~~~~~~~~~~~~~~~~~~~~~~~~~~~~~~~~~~~~~~~~~~~~~~~~~~~~~
1240 N=2: ODD=0: EVEN=0
1250 Z2=4: POLD=0
1260 SUM = FNF(Z1) + FNF(Z2)
1270 DELT = ABS (Z2-Z1)/N
1280 Z =Z1 +DELT
1290 ODD = FNF(Z)
1300 FOR J=1 TO N-1 STEP 2
1310    Z=Z+2*DELT
1320    ODD = FNF(Z) + ODD
1330 NEXT J
1340 PROB = DELT/3*(SUM+4*ODD +2*EVEN)
1350 IF ABS(PROB-POLD)<=.001 THEN GOTO 1400
1360 EVEN =EVEN+ODD
1370 POLD = PROB
1380 N = N*2
1390 GOTO 1270
1400 PROB = PROB+.00003
1410 RETURN
1420 PRINT:PRINT"END OF R-INDEX PROGRAM"
1430 END
```

```
RUN
***************************************************************
*************  WELCOME TO THE R-INDEX CALCULATION  ***********
***************************************************************


Enter the number of categories in the R-index matrix: ? 4

Enter the scores in the top row of the matrix, (include zeros).

SCORE 1 : 6
SCORE 2 : 2
SCORE 3 : 2
SCORE 4 : 0


Enter the scores in the bottom row of the matrix.

SCORE 1 : 2
SCORE 2 : 2
SCORE 3 : 5
SCORE 4 : 7

-----------------------------------------------------------------
TOP ROW (S):  6     2     2     0
BOT ROW (N):  2     2     5     7
-----------------------------------------------------------------

Are all these data values correct? (YES OR NO): ? N

Is the top row correct? (YES OR NO): ? Y

Is the bottom row correct? (YES or NO): ? N


Enter the scores in the bottom row of the matrix.

SCORE 1 : 0
SCORE 2 : 1
SCORE 3 : 2
SCORE 4 : 7

-----------------------------------------------------------------
TOP ROW (S):  6     2     2     0
BOT ROW (N):  0     1     2     7
-----------------------------------------------------------------

Are all these data values correct? (YES OR NO): ? Y

---------------------------------------
THE R-INDEX IS: 0.9550  OR  95.50 %
---------------------------------------


THE RANK SUMS TEST Z VALUE IS:  3.37381

THE PROBABILITY OF GETTING AN R THIS LARGE

ON THE NULL HYPOTHESIS IS:
      ONE-TAILED = 0.00039
      TWO-TAILED = 0.00077


 ENTER 1 TO CONTINUE R-INDEX WITH  4  CATEGORIES
 ENTER 2 TO CONTINUE R-INDEX WITH DIFFERENT NUMBER OF CATEGORIES
 ENTER 3 TO EXIT PROGRAM

     OPTION NUMBER: 3

END OF R-INDEX PROGRAM
Ok
```

REFERENCES

1. **O'Mahony, M., Wong, S.-Y., and Odbert, N.,** Sensory evaluation of Regina freestone peaches treated with low doses of gamma-radiation, *J. Food Sci.,* 50, 1051, 1985.

2. **O'Mahony, M, Buteau, L., Klapman, K., Stavros, I., Alford, J., Leonard, S. J., Heil, J. R., and Wolcott, T. K.,** Sensory evaluation of high vacuum flame sterilized clingstone peaches, using ranking and signal detection measures with minimal cross sensory interference, *J. Food Sci.,* 48, 1626, 1983.

3. **O'Mahony, M., Thompson, B., Davies, M., Gardner, L., Long, D., and Heintz, C.,** Salt taste detection: an R-index approach to P(A) measurements, *Perception,* 8, 497, 1979.

4. **Green, D. M. and Swets, J. A.,** *Signal Detection Theory and Psychophysics,* John Wiley & Sons, New York, 1966.

5. **Amerine, M. A., Pangborn, R. M., and Roessler, E. B,** *Principles of Sensory Evaluation of Food,* Academic Press, New York, 1965, chap. 7.

6. **O'Mahony, M. and Odbert, N.,** A comparison of sensory testing procedures: sequential sensitivity analysis and aspects of taste adaptation, *J. Food Sci.,* 50, 1055, 1985.

7. **O'Mahony, M., Wong, S.-Y., and Odbert, N.,** Sensory difference tests: some rethinking concerning the general rule that more sensitive tests use fewer stimuli, *Lebensm. Wiss. Technol.,* 19, 93, 1985.

8. **Gridgeman, N. T.,** Group size in taste sorting trials, *Food Res.,* 21, 534, 1956.

9. **Harris, H. and Kalmus, H.,** The measurement of taste sensitivity to phenylthiourea (PTC), *Ann. Eugenics,* 15, 24, 1949.

10. **Fallis, N., Lasagna, L., and Tétreault, L.,** Gustatory thresholds in patients with hypertension, *Nature (London),* 196, 74, 1962.

11. **Furchtgott, E. and Willingham, W. W.,** The effect of sleep deprivation upon the thresholds of taste, *Am. J. Psychol.,* 69, 111, 1956.

12. **Schelling, J. L., Tétreault, L., Lasagna, L., and Davis, M.,** Abnormal taste threshold in diabetes, *Lancet,* 1, 508, 1965.

13. **Topinka, I and Sova, J.,** Chutovy prah NaCl, jeho geneticka vazba a mozny vyznam v etiologii hypertenze, *Cosopis Lekaru Cesk,* 106, 689, 1967.

14. **Pokorný, J., Marcin, A., and Davidek, J.,** Comparison of the efficiency of triangle and tetrade tests for discrimination sensory analysis of food, *Die Nahrung,* 25, 561, 1981.

15. **Roessler, E. B., Pangborn, R. M., Sidel, J. L., and Stone, H.,** Expanded statistical tables for estimating significance in paired-preference, paired-difference, duo-trio and triangle tests, *J. Food Sci.,* 43, 940, 1978.

16. **O'Mahony, M.,** *Sensory Evaluation of Food: Statistical Methods and Procedures,* Marcel Dekker, New York, 1986.

17. **O'Mahony, M.,** Salt taste adaptation: the psychophysical effects of adapting solutions and residual stimuli from prior tastings on the taste of sodium chloride, *Perception,* 8, 441, 1979.

18. **O'Mahony, M., Hobson, A., Garvey, J., Davies, M., and Birt, C.,** How many tastes are there for low concentration "sweet" and "sour" stimuli? — threshold implications, *Perception,* 5, 147, 1976.

19. **O'Mahony, M., Kingsley, L., Harji, A., and Davies, M.,** What sensation signals the salt taste threshold?, *Chem. Senses Flavor,* 2, 177, 1976.

20. **O'Mahony, M. and Stevens, J.,** Criterion points and qualitative taste descriptive categories along the taste continuum for citric acid solutions — preliminary study, *IRCS Med. Sci.,* 3, 546, 1975.

21. **O'Mahony, M.,** Some assumptions and difficulties with common statistics, *Food Technol.,* 36, 75, 1982.

22. **McNicol, D.,** *A Primer of Signal Detection Theory,* Unwin, London, 1972, 50.

23. **O'Mahony, M.,** Salt taste sensitivity: a signal detection approach, *Perception,* 1, 459, 1972.

24. **O'Mahony, M.,** Purity effects and distilled water taste, *Nature (London),* 240, 489, 1972.

25. **Brown, J.,** Recognition assessed by rating and ranking, *Br. J. Psychol.,* 65, 13, 1974.

26. **Elliott, P. B.,** tables of d', in *Signal Detection and Recognition by Human Observers: Contemporary Readings,* Swets, J. A., Ed., John Wiley & Sons, New York, 1964, 651.

27. **Frijters, J. E. R., Kooistra, A., and Vereijken, P. F. G.,** Tables of d' for the triangular method and the 3-AFC signal detection procedure, *Percept. Psychophys.,* 27, 176, 1980.

28. **O'Mahony, M., Atassi-Sheldon, S., Wong, J., Klapman-Baker, K., and Wong, S.-Y.,** Salt taste sensitivity and stimulus volume: sips and drops. Some implications for the Henkin taste test, *Perception,* 13, 725, 1985.

29. **O'Mahony, M., Klapman, K., Wong, J., and Atassi, S.,** Salt taste sensitivity and stimulus volume: effect of stimulus residuals, *Perception,* 11, 347, 1982.

30. **O'Mahony, M., Atassi-Sheldon, S., Rothman, L., and Murphy-Ellison, T.,** Relative singularity/mixedness judgments for selected taste stimuli, *Physiol. Behav.,* 31, 749, 1983.

31. **O'Mahony, M., Garske, S., and Klapman, K.,** Rating and ranking procedures for short-cut signal detection multiple difference tests, *J. Food Sci.,* 45, 392, 1980.

32. **Leach, C.,** *Introduction to Statistics,* John Wiley & Sons, New York, 1979, 49.

33. **Runyon, R. P. and Haber, A.,** *Fundamentals of Behavioral Statistics,* Addison-Wesley, Reading, Mass., 1967, 250.
34. **O'Mahony, M., Wong, S.-Y., and Odbert, N.,** Sensory evaluation of navel oranges treated with low doses of gamma-radiation, *J. Food Sci.,* 50, 639, 1985.
35. **O'Mahony, M., Kulp, J., and Wheeler, L.,** Sensory detection of off-flavors in milk, incorporating short-cut signal detection measures, *J. Dairy Sci.,* 62, 1857, 1979.
36. **O'Mahony, M., Wong, S.-Y., and Odbert, N.,** Initial sensory evaluation of bing cherries treated with low doses of gamma-radiation, *J. Food Sci.,* 50, 1048, 1985.

Chapter 9

USES AND ABUSES OF CATEGORY SCALES IN SENSORY MEASUREMENT

Dwight R. Riskey

TABLE OF CONTENTS

I.   Introduction and Objectives 178

II.   The History of Category Scales 178
      A.   Weber 178
      B.   Fechner 178
      C.   Thurstone 179
      D.   Direct Scaling 180

III. Analogy to Fundamental Measurement 181

IV.   Contextual Influences 181
      A.   Range of Stimuli 181
      B.   Frequency Effects 182
      C.   Range-Frequency Model 183
      D.   Number of Categories 186
      E.   Number of Stimuli 187
      F.   Sequential Effects 187
      G.   Ratings of Stimulus Acceptability 188

V.   Summary and Practical Implications 189

References 190

I. INTRODUCTION AND OBJECTIVES

Category ratings are heavily used by social scientists for a variety of reasons, including assessment of quality of life, public opinion, and simple perceptions. During the last several decades, they have gained in popularity as a device for assessing consumer preferences and for guiding efforts in the development of new products. They have been scorned as invalid by some and touted as direct pipelines to the psyche by others. From time to time, however, they have been misinterpreted or misused by almost everybody. This chapter attempts to demystify category ratings by outlining the history of their development and by describing their patterns of behavior. Special emphasis is placed on common misuses of category scales and how they can best be used to measure human perceptions, impressions, and values.

II. THE HISTORY OF CATEGORY SCALES

A. Weber

The development of category scaling as a device for measuring psychological or sensory impressions can be tracked through the history of psychophysics to Weber's Law. Weber, in his studies of kinesthesis, discovered that the amount of weight necessary to elicit a just noticeable difference varies depending upon the weights used; heavy weights require greater change in physical weight for difference detection than light weights. When carefully measured, the ratio between any weight and a weight just noticeably different was observed to be approximately constant no matter what absolute weights were considered. Specifically, the ratio of the amount of weight producing a just detectable difference ($\Delta\Phi$) to the weight of a standard (Φ) remained constant for different weights of the standard:

$$\frac{\Delta\phi}{\Delta} = k$$

Of course, different constants (k) would apply for different modalities.[1] Since the Weber fraction indicates the proportion by which any stimulus continuum must change in physical value to produce a changed psychological impression, the Weber fraction indicates the relative sensitivities of the various sensory dimensions. For example, a sensory system such as brightness vision requiring only about a 2% change in stimulus intensity to produce a just noticeable difference (JND) is more sensitive than the taste system, which requires as much as a 33% concentration change for detection.

Ostensibly, Weber's Law seems to provide an elegant description of the relationship between the physical world and psychological impressions. However, as Guilford[2] points out, Weber's fraction actually relates two physical measures, a standard and its just noticeably different counterpart. The Weber fraction allows us to predict the intensity of a stimulus just noticeably different from a standard, but does not directly allow the prediction of psychological or sensory magnitude.

B. Fechner

Fechner saw in Weber's principle, however, that JNDs provide a critical psychological measuring unit which could serve as the basis for a true scale of sensory impressions. Fechner reasoned that since a JND is a psychologically determined minimum, each JND must be equal in psychological value. In order to scale the psychological magnitude of a particular physical stimulus, the number of psychological JNDs above absolute threshold can therefore simply be added.[3] For example, if absolute threshold is arbitrarily set at 12 for the hypothetical case shown in Figure 1 and the Weber fraction is 0.25, the physical stimulus increases by

Scaling according to Fechner's Logarithmic Law. (k = .25)

Psychological JND Steps	Stimulus Value	Log Stimulus	Increment of Log
0	12 (abs. threshold)*	1.079	—
1	15 (12 + 12/4 = 15)	1.176	0.97
2	18.75 (15 + 15/4 = 18.75)	1.273	0.97
3	23.44	1.370	0.97
4	29.30	1.467	0.97
5	36.64	1.564	0.97
6	45.79	1.661	0.97
7	57.24	1.758	0.97

FIGURE 1.   Scaling according to Fechner's Logarithmic Law, K = 0.25. (Adapted from Engen, T., in *Experimental Psychology*, Vol. 1, Kling, J. W. and Riggs, L. A., Eds., Holt, Rhinehart & Winston, New York, 1972.)

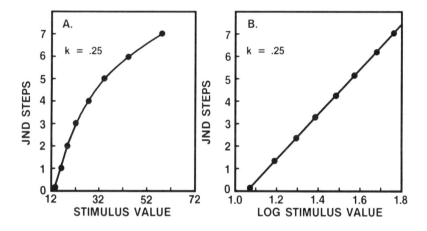

FIGURE 2.   JND steps as a function of stimulus value. Panel A plots stimulus value on a linear scale; panel B plots stimulus value on a logarithmic scale.

the same ratio or percentage with each JND. And, of course, in logarithmic terms, the increment of the log of the stimulus value will be equal for each step.

The data are plotted in Figure 2, Panel A on linear scales and Panel B on semilogarithmic coordinates. This relationship between JND steps and stimulus value has become known as Fechner's Psychophysical Law. In simplest terms, it states that equal ratios in stimulus steps (shown here by equal steps on a logarithmic scale) produce equal psychological differences (or JNDs).

C. Thurstone

Thurstone then demonstrated that measuring JNDs to generate such psychological scales, a time-consuming and awkward process, is unnecessary. His Law of Comparative Judgment[4] pointed out that the degree to which the psychological impressions of any two stimuli in a set overlap (i.e., the amount of confusion between the stimuli) is enough information to infer the psychological distances among the various stimuli.

As shown in Figure 3, Thurstone contended that every stimulus on any given exposure elicited one of a possible distribution of perceptions. These different perceptions for the

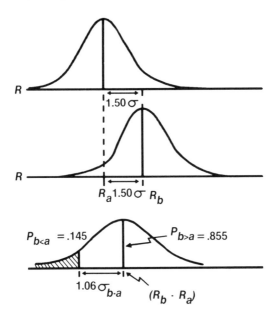

FIGURE 3.   Hypothetical discriminal dispersions for two stimuli, R_a and R_b. The distribution of differences shown in the bottom panels allows inferences about the number of perceptual dispersion units separating the two stimuli. (Adapted from Guilford, J. P., *Psychometric Methods*, McGraw-Hill, New York, 1954.)

same stimulus are due to all sorts of internal noise or random factors in the perceptual system. At any given moment, a perceptual stimulus may have a larger or smaller perceptual effect on the respondent. As a result, if the two stimuli of interest, R_a and R_b, are not too far apart, there will be some overlap or confusion between them. If the proportion of times one is judged more intense (or more sweet or more acceptable) is known, a distribution of differences can be constructed from which the number of dispersion units separating the two stimuli can be inferred. These dispersion units are analogous to a JND scale but bring perceptual measurement a step closer to what we know as modern category scaling.

D. Direct Scaling

Note how similar Thurnstonian scaling is to a categorization operation. Essentially, Thurstonian scales derive from questions such as, "Which is larger or sweeter or more acceptable?" It was soon common practice to ask respondents for more information. "Which product is sweeter, and by how much?" "Sort the stimuli into the following categories: very sweet, slightly sweet, and not sweet at all." Procedures were developed to turn this type of categorical data into Thurstonian scales. One approach produced about the same results as the other, but since the categorical approach was so much easier and more efficient, it gained in popularity.

At this point, a really big empirical and historical leap was taken. Thurstonian scales generated in the "indirect" way described above were compared to simple face value categorical judgments. A close relationship between the two approaches was observed (i.e., the "indirect" vs. "direct" tended to produce the same results) and category ratings in their current popular form were off and running.[5]

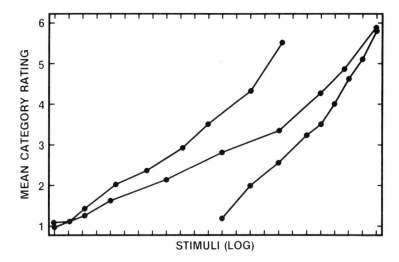

FIGURE 4. Ratings of the sizes of squares in three stimulus sets, each having different ranges. (From Parducci, A. and Perrett, L. F., *J. Exp. Psychol. Monogr.*, 89, 427, 1971. Copyright 1971 by the American Psychological Association. Reprinted by permission of the publisher and author.)

III. ANALOGY TO FUNDAMENTAL MEASUREMENT

Despite their heavy and growing use, category ratings are frequently misinterpreted and misused. Almost all of this misunderstanding is rooted in the inevitable analogy to fundamental measurement. An implicit belief exists that, as in physical measurement, a particular scale value or judgment must always be associated with a particular physical stimulus. This belief lies at the very heart of what is meant by reliability and validity in measuring physical characteristics of objects. A scale that provides one numerical weight one day and another weight the next for the same object is pretty quickly fixed or discarded.

For scales of psychological measurement, however, this type of behavior is common. The same physical stimulus evaluated by two individuals or by the same individual in two different situations very often does not evoke the same judgmental response. A "warm" day in June may be quite different in an absolute sense from a "warm" day in February. A food that tastes "very good" in one context may be unacceptable in another. These examples demonstrate that judgments, unlike physical measurements, tend not to be made against a set of fixed standards. Judgments tend to be relative in nature. They depend upon a number of situational factors and it follows that for a scale to be useful in measuring psychological impressions, it must reflect these situational influences. Moreover, understanding these influences is critical in the proper interpretation of the scales.

IV. CONTEXTUAL INFLUENCES

A. Range of Stimuli

A particular stimulus will be judged differently depending upon the range of stimuli in the set within which it is presented. The data in Figure 4 show a classic example of this effect.[6,7] In this particular case, three groups of respondents rated sizes of squares. Each group rated one of three sets of squares, each with different end points. The slope is steepest for the narrow range, less steep for the wider range. The respondents perform as if the ends of a rubber ruler were being applied to the end points of the stimulus range. The very same stimulus can thus be judged "small" or "large" depending upon the range of stimuli within which it is presented.

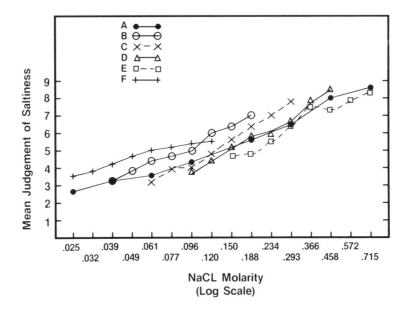

FIGURE 5.   Mean saltiness ratings of solutions as a function of NaCl concentration.
Each curve represents judgments of stimulus sets with a different range.

Figure 5 shows similar data for the effects of range of salt concentrations on judgments of saltiness of solutions. Six different groups of 60 respondents produced each curve.[8] Generally, narrower stimulus ranges produce steeper functions; also, a narrow range at the low end of the overall range tends to generate a judgment of saltier taste for a particular solution than an equally narrow range towards the top end of the overall range.

Figure 5 points out another aspect of category ratings that deserves mention. Some investigators have observed that in certain circumstances, the end points of the category rating scale are used less frequently than other categories and that this may produce ceiling or floor effects. Some slight evidence for such effects may be apparent in Figure 5 for the widest range. One way to deal with these effects is to anchor the extremes of the scale with stimuli not to be included in the set.[9,10] This tends to make all these range conditions move closer together, thereby minimizing the range effect. However, these anchors should not be too extreme. If they are, they will tend to not be included as part of the psychological set, will be discounted by respondents, and thereby will have reduced impact on minimizing the effects of stimulus range.[5,8]

B. Frequency Effects

Another way in which category ratings can produce nonlinearly related scales for the same set of stimuli is through the effects of stimulus frequency or spacing. Figure 6 from Parducci and Perret[6] shows both of these effects very clearly. The upper abscissa shows the relative frequency of presentation of various stimulus intensities (sizes of squares for this particular investigation); the open points connected by the dashed line show the empirical judgments for the various stimuli. The judgment function is steepest where the stimuli are most frequent, flatter where the stimuli are less frequent. Another way to produce these types of effects, through manipulations of the stimulus spacings, is also shown in Figure 6. The bottom curve shows data generated from an experiment in which the stimuli were spaced nearer together towards the top end of the range tested. As one would expect, the judgment curve is steepest at the top end of the stimulus range where the stimuli are most bunched together. Note that the ranges of the stimuli in both stimulus sets are identical.

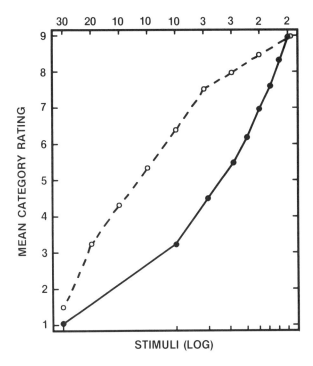

FIGURE 6. Ratings of stimuli presented with unequal spacing (lower abscissa, solid curve) or unequal frequencies (upper abscissa, dashed curve). (From Parducci, A. and Perrett, L. F., *J. Exp. Psychol. Monogr.*, 89, 427, 1971. Copyright 1971 by the American Psychological Association. Reprinted by permission of the publisher and author.)

Similar effects of stimulus spacing have been demonstrated by Stevens and Galanter.[11,12] Figure 7 plots results of an experiment in which two different stimulus spacings covering the same stimulus range were investigated. Stimulus spacings represented on the lower abscissa are shown as solid points; open points reflect mean judgments of spacings represented on the upper abscissa. Category ratings for the two spacings are dramatically different. Slopes of the judgment functions are steepest where stimuli are most bunched together. Note that ordinal relationships between stimuli within a stimulus set are maintained. This results in a judgment function reflecting overall greater judged stimulus size for sets bunched towards the bottom of the range.

A final example of the effects of stimulus frequencies in the domain of taste is shown in Figure 8. Riskey[13] asked respondents to judge the saltiness of soups prepared with various concentrations of salt. One group of subjects tasted a series of soups in which the lowest concentrations were most frequent, a positively skewed distribution. Another group judged a series in which the highest concentrations were most frequent, a negatively skewed distribution. A third and a fourth group judged two different symmetrical distributions of concentrations. The data are consistent with the previously described frequency effects. It is as if respondents are trying to make greater discriminations between stimuli that are bunched together or more similar. A more operational or behavioristic view is that respondents are trying to use each of the categories in the rating scale equally often.

C. Range-Frequency Model

Parducci[14] has proposed a mathematical model to account for the pervasive effects of stimulus range and frequency in category ratings. The model states that category ratings are

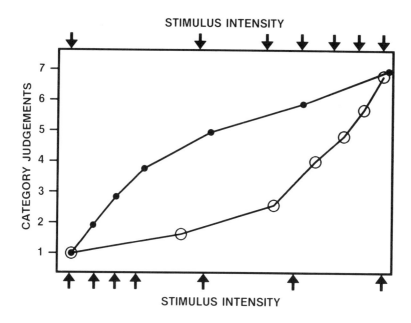

FIGURE 7.   Category ratings for two different stimulus spacings. (From Galanter, E., *Textbook of Elementary Psychology*, Holt, Rhinehart & Winston, 1966. With permission.)

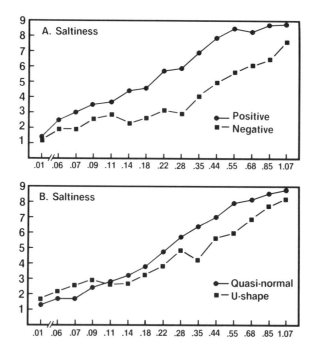

FIGURE 8.   Mean judgments of saltiness for 15 concentrations of NaCl presented in four different frequency distributions. (Adapted from Riskey, D. R., *Selected Sensory Methods: Problems and Approaches to Measuring Hedonics*, Kuznicki, J. T., Johnson, R. A., and Ruthiewic, A. F., Eds., American Society for Testing and Materials, Philadelphia, 1982. With permission.)

The Range-Frequency Model of Category Judgements — A Simple Example:

The Range Principle . . . Subjects assign ratings to equal
sections of the stimulus range.

The Frequency Principle . . . Subjects assign equal numbers
of stimuli to each category.

Assume the simplest possible case. Two categories of judgement. . .
"Large", and "Small". The distribution of stimulus frequencies
is positively skewed.

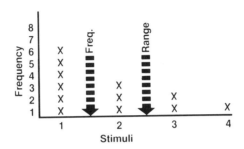

Range Principle . . . The Category limen is the Median.
(i.e., 1 = "Small"; 2-4 = "Large".)

Frequency Principle . . . The category limen is the Midpoint.
(i.e., 1 & 2 = "Small"; 3 & 4 = "Large".)

Range-Frequency Prediction . . . R-F limen is "2".
Stimulus 1 = "Small"
Stimulus 2 = ½ "Small"
½ "Large"
Stimulus 3 = "Large"
Stimulus 4 = "Large"

FIGURE 9.   A simple example of category rating predictions based on Parducci's range-frequency model.

a simple compromise between a range principle, the tendency to assign categories to equal sections of the stimulus range, and the frequency principle, the tendency to assign equal numbers of stimuli to each category.

The simplest possible case involving only two categories "Large" and "Small", is summarized in Figure 9. The example assumes that the frequency distribution of the stimuli is positively skewed. The problem is to predict the stimulus value at which respondents will draw the limen between "Large" and "Small." According to the range principle, the limen should go at the midpoint such that it divides the range of stimuli into two equal sections. According to the frequency principle, the limen should go at the median such that equal numbers of stimuli fall into each category. The range-frequency prediction is that the actual dividing point will fall midway between the median and midpoint. The smallest stimulus will be called "small"; the largest two stimuli will be called "large". The remaining stimulus has an equal chance of being assigned to either category. Of course, with greater numbers of categories and stimuli the situation is more complex. The principles, however, can be applied to any number of categories and stimuli.

The range-frequency model has done extremely well in predicting category judgments, providing we know range and frequency information about the stimuli to be evaluated (i.e., providing we know something about context of the situation under investigation).[6,7,14,16]

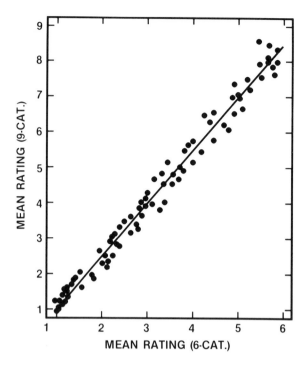

FIGURE 10.   Linear relationship between judgments from category rating scales containing different numbers of categories. (From Parducci, A. and Perrett, L. F., *J. Exp. Psychol. Monogr.*, 89, 427, 1971. Copyright 1971 by the American Psychological Association. Reprinted by permission of the publisher and authors.)

D. Number of Categories

Another feature of category ratings that is constantly under debate is the optimum number of categories to include in a scale. Does number of categories have any effect on the results produced? The most common argument for using many categories is largely intuitive. Response will be less constrained and more veridical if response choices are less limited. The data, however, do not support this view. Figure 10 plots the results of an investigation by Parducci[15] in which judgments of the sizes of squares were made in two conditions identical in every way except number of categories. One condition involved the use of a nine-point category scale, while the other included only six categories. Judgments in the two conditions are clearly linearly related.

Unfortunately, the situation is not quite so simple as these data make it appear. The magnitude of the effects of stimulus range and frequency, as described above, is dramatically affected by the number of categories available for use.[15,16] Figure 11 shows that contextual effects associated with skews in the distribution of stimuli in the set (sizes of squares) are smaller when larger numbers of categories are employed. The dashed curves represent a negatively skewed distribution of stimulus frequencies while solid curves represent positively skewed distributions. The really big drops occur between 2, 3, and 4 categories, but the decline is systematic all the way through scales having 20 or even 100 categories.

The "open" condition is particularly interesting. Here respondents were allowed to generate as few or as many categories as they felt were appropriate. Virtually all of the respondents chose to use from 3 to 12 categories and the sizes of the frequency effects are appropriate to the number of categories they used.

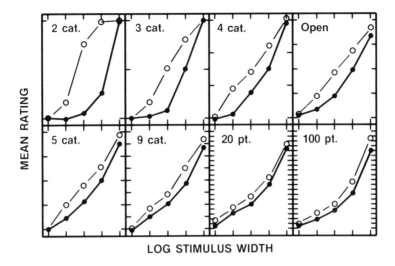

FIGURE 11.   Effects of skewing the stimulus distribution vary inversely with number of categories. (From Parducci, A. and Wedell, D. H., *J. Exp. Psychol. Hum. Percept. Performance*, 12(4), 496, 1986. With permission.)

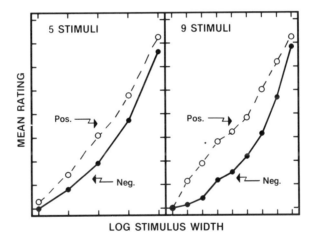

FIGURE 12.   Effects of skewing the stimulus distribution vary directly with increasing numbers of stimuli. (From Parducci A. and Wedell, D. H., *J. Exp. Psychol. Hum. Percept. Performance*, 12(4), 496, 1986. With permission.)

E. Number of Stimuli

There is also evidence that the number of stimuli in a set can affect the size of the frequency effect.[17] Figure 12 plots experimental conditions which are identical in every way except the number of stimuli making up the range of stimuli. Note the difference in the degree to which the two conditions are affected by skewing of the stimulus distributions. Large numbers of stimuli spread across a given range produce larger frequency effects than small numbers of stimuli.

F. Sequential Effects

There is strong evidence for sequential effects in category ratings. Figure 13 shows a graph adapted from Holland and Lockhead[17] showing that the first stimulus following a

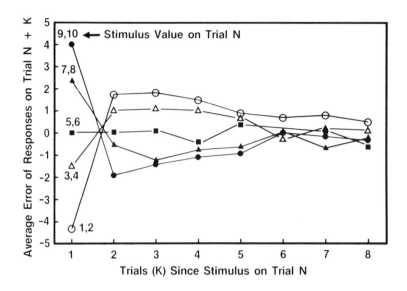

FIGURE 13. Effects of a stimulus on a given trial on responses across the next eight trials. (From Holland, M. K. and Lockhead, G. R., *Perception and Psychophysics, 3,* 409, 1968. With permission.)

stimulus that is rated high will be rated higher than it otherwise would have been. The first stimulus after a low stimulus is conversely rated lower than it otherwise would be. Subsequent stimuli instead show a contrast effect relative to that initial stimulus. Moreover, that contrast effect fades in magnitude with each passing trial. Initially, then, there tends to be an assimilation effect (stimuli tend to be judged more like the preceding stimulus), but a contrast effect (stimuli tend to be judged less like the preceding stimulus) appears with passing trials.

G. Ratings of Stimulus Acceptability

Ratings of stimulus acceptability represent a special case with respect to contextual effects because, for many characteristics of stimuli, hedonic impressions are nonmonotonically related to physical intensity. In the case of foods, for example, increasing the concentration of an ingredient such as salt or sugar contributes to its acceptability to a certain peak level beyond which acceptability typically declines.[18-24] Effects of manipulating the range or frequency of stimuli within the set are not obvious and, because of this nonmonotonicity, cannot be directly derived from Parducci's range-frequency model. The observed effects, however, are consistent with the relational nature of judgments upon which the range-frequency model is based.

Figure 14 plots results of an experiment by Riskey et al.[25] in which respondents judged sweetness and pleasantness of beverages prepared with various concentrations of sugar. The upper panel shows effects of three different distributions of stimulus frequencies on judgments of sweetness. The familiar contextual effects are apparent. Judgmental functions are steeper where stimuli are more frequent. For example, a given concentration is judged much sweeter in the positively skewed context, in which beverages containing lower concentrations are more frequent, than in the negatively skewed context containing mostly higher concentrations. Judgments of the quasi-normal distribution fall primarily between the two skewed conditions.

The lower panel shows that low and high concentrations are less pleasant than concentrations falling between the two extremes. The most interesting aspect of the data is that the concentration judged maximally acceptable differs as a function of judgmental context. The

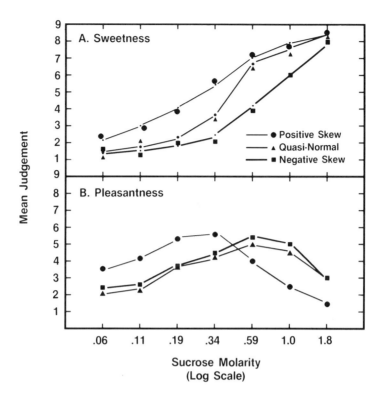

FIGURE 14.   Mean ratings of sweetness and pleasantness for seven concentrations of sucrose presented in those different frequency distributions. (From Riskey, D. R., Parducci, A., and Beauchamp, G. K., *Percept. Psychophys.*, 26(3), 171, 1979. With permission.)

positively skewed context, in which stimuli are judged sweeter, produces maximum acceptability at lower concentrations than the negatively skewed distribution.

Such shifts in the point of maximum pleasantness are not the result of a simple sensory adaptation.[13] Instead, a kind of psychological recalibration seems to be associated with changes in judgmental context. The same stimulus takes on different hedonic value depending upon the other stimuli to which it is compared.

Recent investigations of the process by which people develop new taste preferences have demonstrated that eating experiences such as low-sodium diets may evoke preference changes analogous to contextual effects established in a laboratory. When placed on low-sodium diets, normal students were found to shift their taste preferences toward foods containing lower levels of salt.[26]

V. SUMMARY AND PRACTICAL IMPLICATIONS

Measurement has been defined as the process of assigning numerical values to objects or events according to rules. According to this definition, category ratings are valid measures. As this chapter has shown, the rules of psychological scaling are somewhat different from, and perhaps more complex than, the rules for physical measurement. Measurement implies to most people the notion that a given stimulus must always correspond to a given number. An important characteristic of psychological measurement, however, is that although we often describe our judgments in what seem to be absolute terms (e.g., excellent, average, or awful), they tend to be relative rather than absolute in nature. No single standard for comparison exists. Instead, the judgmental scheme is constantly in a state of flux and the frame of reference for a given judgment is often difficult to specify.

It could be argued that product evaluators should welcome methods of scaling which allow situational factors to be a part of the judgmental scheme. Instead, the situational dependency and relativity of human judgments are typically viewed as roadblocks to the process of uncovering the true essence of a product. Part of the problem is that the main topic of interest in sensory evaluation is the product, not the respondent. Although situational dependency of human judgments is seen as interesting, it limits the degree to which product judgments can be compared across time and across investigations. Knowing that a product received an average judgment of 7.5 on a nine point category scale, for example, is of little value unless the context for judgment and relevant measurement parameters are specified. An objective of the present chapter was therefore to delineate those measurement parameters and the manner in which they influence judgments.

Contextual effects such as those described in this chapter are not unique to category ratings.[27] Scaling procedures from Fechner forward have been affected by at least some of these factors. Psychological scales in sensory evaluation, because they do not conform to the laws of fundamental measurement, have value in direct proportion to the degree that their response to situational dependencies can be understood. Category ratings have the advantage that the manner in which they are affected is well understood. For many of the contextual variables discussed in this chapter, their effects on category ratings can be predicted quantitatively or at least directionally. One does not have this luxury with many other types of scales.

REFERENCES

1. **Geldard, F. A.,** *Fundamentals of Psychology,* John Wiley & Sons, New York, 1962.
2. **Guilford, J. P.,** *Psychometric Methods,* McGraw-Hill, New York, 1954.
3. **Engen, T.,** Psychophysics II, scaling methods, in *Experimental Psychology,* Vol. 1, Kling, J. W. and Riggs, L. A., Eds., Holt, Rinehart & Winston, New York, 1972, chap. 3.
4. **Thurstone, L. L.,** *Measurements of Values,* Univ. of Chicago Press, 1959.
5. **Bock, R. D. and Jones, L. V.,** *The Measurement and Prediction of Judgment and Device,* Holden-Day, San Francisco, 1968.
6. **Parducci, A. and Perrett, L. F.,** Category rating scales: effects of relative spacing and frequency, *Exp. Psychol. Monogr.,* 89, 427, 1971.
7. **Parducci, A.,** Contextual effects: a range-frequency analysis, in *Handbook of Perception,* Vol. 2, Carterette, E. C. and Friedman, M. P., Eds., Academic Press, New York, 1974, chap. 5.
8. **Riskey, D. R.,** unpublished data, 1981.
9. **Jones, N. F.,** Overview of psychophysical scaling methods, in *Handbook of Perception,* Vol. 2, Carterette, E. C. and Friedman, M. P., Eds., Academic Press, New York, 1974, chap. 10.
10. **Sarris, V.,** Adaptation-level theory: two critical experiments on Helson's weighted average model, *Am. J. of Psychol.,* 80, 331, 1967.
11. **Stevens, S. S. and Galanter, E. H.,** Ratio scales and category scales for a dozen perceptual continua, *J. Exp. Psychol.,* 54, 377, 1957.
12. **Galanter, E.,** *Textbook of Elementary Psychology,* Holt, Rinehart & Winston, San Francisco, 1966.
13. **Riskey, D. R.,** Effects of context and interstimulus procedures in judgments of saltiness and pleasantness, in *Selected Sensory Methods: Problems and Approaches to Measuring Hedonics,* Special Technical Publication 773, Kuznicki, J. T., Johnson, R. A., and Rutkiewic, A. F., Eds., American Society for Testing and Materials, Philadelphia, 1982, 71.
14. **Parducci, A.,** Category judgment: a range-frequency model, *Psychol. Rev.,* 72, 407, 1965.
15. **Parducci, A.,** Category rating: Still more contextual effects, in *Social Attitudes and Psychophysical Measurement,* Wegener, B., Ed., Erlbaum, Hillsdale, N.J., 1982, chap. 3.
16. **Parducci, A. and Wedell, D. H.,** The category effect with rating scales: number of categories, number of stimuli, and method of presentation, *J. Exp. Psychol. Hum. Percept. Performance,* 12(4), 496, 1986.

17. **Holland, M. K. and Lockhead, G. R.,** Sequential effects in absolute judgments of loudness, *Percept. and Psychophys.,* 3, 409, 1968.

18. **Kocher, E. C. and Fisher, G. L.,** Subjective intensity and taste preference, *Percept. Mot. Skills,* 28, 735, 1969.

19. **Moskowitz, H. R.,** The sweetness and pleasantness of sugars, *Am. J. Psychol.,* 84, 387, 1971.

20. **Moskowitz, H. R.,** Sensations, measurement and pleasantness: confessions of a latent introspectionist, in Taste and Development: The Genesis of Sweet Preference, Weiffenbach, J. M., Ed., U.S. Department of Health, Education & Welfare, Bethedsa, MD, 1977, chap. 20.

21. **Moskowitz, H. R.,** Intensity and hedonic functions for chemosensory stimuli, in *The Chemical Senses and Nutrition,* Kare, M. R. and Maller, O., Eds., Academic Press, New York, 1977, chap. 4.

22. **Moskowitz, H. R., Kluter, R. A., Westerling, J., and Jacobs, J. L.,** Sugar sweetness and pleasantness: evidence for different psychological laws, *Science,* 184, 583, 1974.

23. **Moskowitz, H. R., Kumraiah, V., Sharma, K. N., Jacobs, H. L., and Sharma, S. D.,** Cross-cultural differences in simple taste preferences, *Science,* 190, 1217, 1975.

24. **Pangborn, R. M.,** Individual variations in affective responses to taste stimuli, *Psychonomic Sci.,* 21, 125, 1970.

25. **Riskey, D. R., Parducci, A., and Beauchamp, G. K.,** Effects of context in judgments of sweetness and pleasantness, *Percept. Psychophys.,* 26(3), 171, 1979.

26. **Bertino, M., Beauchamp, G. K., and Engelman, K.,** Long-term reduction in dietary sodium alters the taste of salt, *Am. J. Clin. Nutr.,* 36, 1134, 1982.

27. **Poultan, E. C.,** The new psychophysics: six models for magnitude estimation, *Psychol. Bull.* 69, 1, 1968.

Chapter 10

MAGNITUDE ESTIMATION: SCIENTIFIC BACKGROUND AND USE IN SENSORY ANALYSIS

Howard R. Moskowitz and Barry E. Jacobs

TABLE OF CONTENTS

I.   Introduction 194
     A.   The Early (Pre-1950s) Studies 195
     B.   Absolute Number Matching or Magnitude Estimation 196
     C.   Scientific Studies Using the Magnitude Estimation Procedure 197
     D.   Methodological Studies 198
     E.   Developing Functional Relations Between Stimuli and
         Perceptions 199
     F.   The Power Law as the Organizing Principle 199

II.   A Magnitude Estimation Experiment of Perceived Odor Intensity
     of Propanol 201
     A.   Selection of Stimuli 201
     B.   Instructions 202
     C.   Orientation in Scaling 202
     D.   Presentation of the Stimuli 203
     E.   Data Normalization — Bringing the Ratings to a Common
         Scale Across Panelists 204
     F.   Other Types of Analysis and Experimental Designs 205

III.   Magnitude Estimation of Liking 207
     A.   The Inverted U Curve as a General Model for Liking Functions 207
     B.   Unipolar vs. Bipolar Scales to Measure Liking 208
     C.   Representative Hedonic Curves Using Magnitude Estimation 209

IV.   Out of the Laboratory — Into the World of Product Testing 209
     A.   Background 209
     B.   Normalizing Magnitude Estimates by External Verbal Standards 212
     C.   Averaging the Magnitude Estimates 213

V.   Using the Magnitude Estimation Procedure for a Product Test 215
     A.   Background 215
     B.   Stimuli 215
     C.   Panel Composition and Activities 215
     D.   Activities During the Evaluations 216
     E.   Analysis of the Ratings 216
     F.   Ratings of the Self-Designed Ideal Product 216
     G.   Relating Sensory Attributes to Liking 217
     H.   Measuring Relative Importance of Attributes by Magnitude
         Estimation 217

VI.   Specific Benefits of Using Magnitude Estimation 219
       A.   Scale Quality 219
       B.   Applicability in Nonnumerical Measurement 219
       C.   Sensitivity to Differences 220
       D.   Improved Ability to Generate Relations Between Variables 220

VII.   Limits of the Magnitude Estimation Method 221
       A.   Orientation Time 221
       B.   Analysis Issues 221

VIII. An Overview 222

References 222

I. INTRODUCTION

In 1946, S. S. Stevens,[1] founder of modern-day psychophysics, suggested that all measurement scales can be assigned to one of four types: nominal, ordinal, interval, or ratio. The ratio scale represents the most desirable form of measurement because it allows researchers to measure quantities in a manner most similar to the way physicists measure physical magnitudes. Table 1 compares the properties of the various scale types and shows the allowable transformations of each scale.

Experimental psychologists interested in the mechanisms of perception became quite interested in the powerful properties of ratio scales. Experience tells us that as we add sugar to coffee, we perceive the coffee as sweeter. We can measure the number of teaspoons of sugar we add to coffee, but what about the perceived sweetness? Does doubling the amount of sugar double sweetness? Does it more than double sweetness? Does it increase sweetness, but to an impression of far less than twice as strong?

The foregoing question has relevance in both basic and applied research. Basic researchers are interested in mechanisms and "laws" of perception. They seek regularities of nature. Researchers measure sensory responses on a scale which has validated ratio properties. In this way, they construct an equation which demonstrates how ratios of sensory magnitudes correspond to ratios of physical magnitudes. (Without a ratio scale, such equations lose much of their power.) Applied researchers also want to develop scales of perception for sensory magnitudes vs. physical stimuli. A perfumer wants to know whether doubling the concentration of a fragrance more than doubles, doubles, or increases but less than doubles the perceived fragrance intensity. The selection of a fragrance submission might well depend upon the sensory intensity function which it generates vs. concentration.

For production of products, scaling is also important. The quality control engineer, wanting to ensure that each production batch of products retains its sensory integrity, has to know whether deviations of physical magnitudes correspond to small, moderate, or large deviations in sensory intensity. For example, small changes in the objectively measured "grit" value of toothpaste may provoke large changes in perceived "grittiness" and consumer complaints. The same percentage change in the physical viscosity of a lotion (e.g., 15% change, up or down) might be hardly noticeable to consumers as a change in perceived thickness. The quality control engineer would like to develop ratio scales of perception in order to learn which physical changes are most able to provoke large sensory responses.

Table 1
FOUR SCALES OF MEASUREMENT[1]

Scale	Operations performed	Permissible transformations	Some appropriate statistics
Nominal	Identify and classify	Substitution of any number for any other number	Number of cases Mode Contingency correlation
Ordinal	Rank order	Any change that preserves order	Median Percentiles Rank-order correlation
Interval	Find distances or differences	Multiplication by a constant Addition of a constant	Mean Standard deviation Product-moment correlation
Ratio	Find ratios, fractions, or multiples	Multiplication by a constant	Geometric mean Percent variability

A. The Early (Pre-1950s) Studies

Ratio scales are well known in physics. The kelvin (K), or absolute, scale of temperature is a good example of the ratio scale. Interval scales of temperature (with no defined 0 points) are typified by the Fahrenheit and Celsius scales. Ratio scales are much more common. Physicists measure mass, length, time, energy, etc. on ratio scales. For the physical scientist, it is unusual and noteworthy to use a scale which does not have valid ratio properties.

Psychologists, especially those involved in perceptual research, have had a more difficult time developing ratio scales. Many researchers abandoned hope for a valid ratio scale and instead used category scales. These scales possess a fixed number of scale points with defined endpoints (a highest and lowest scale value). Examples of these category scales appear in Table 2.

During the 1930s, researchers in the psychology department at Harvard University developed techniques to erect valid ratio scales, using approaches such as fractionation and multiplication. As an example of the approach, consider the method described below to develop a ratio scale of taste intensity for sugar sweetness:

Step 1 — Create a set of sugar solutions increasing in physical concentration by modest changes (e.g., 15%). The researcher creates the solution set and proceeds to Step 2.

Step 2 — Present the panelist with a starting stimulus sugar solution, usually in the middle of the range. Call that stimulus "1". Let the panelist taste the solution to get an idea of its sweetness.

Step 3 — Instruct the panelist to find a solution which tastes "2" or twice as strong as the reference stimulus. The panelist must taste the various test solutions prepared in step 1 and locate the particular stimulus tasting as close to a "2" as possible.

Step 4 — Continue the procedure. The panelist must find a stimulus which tastes as sweet as a "4" (or twice as sweet as the stimulus rated "2").

Step 5 — Continue, until the panelist has selected an array of sugar solutions exhibiting the desired ratio properties, and whose sensory sweetness values correspond to the values assigned by the researcher. This approach generates a set of numbers (scale values) having presumed ratio scale properties, and a set of corresponding sugar solutions, one for each scale value.

Step 6 — Develop an equation which relates the rating (on the ratio scale) to the concentration of the sugar solution.

The foregoing procedure[2] was followed in a number of cases to generate ratio scales for a variety of sense continua including loudness,[3] taste,[4] etc. The method is tedious, but it

Table 2
EXAMPLE OF
CATEGORY SCALES
USED TO MEASURE
INTENSITY AND LIKING

Scales of perceived intensity
Extremely strong
Very strong
Strong
Moderate
Weak
Cannot perceive

Extremely strong
Strong
Mild
Extremely mild

Strong
Moderate
Weak
Cannot perceive

Scales of liking
Excellent
Very good
Good
Fair
Poor

Like extremely
Like very much
Like moderately
Like slightly
Neither like nor dislike
Dislike slightly
Dislike moderately
Dislike very much
Dislike extremely

generates reliable data after the researcher averages together the results from a relatively small group of panelists. (Approximately ten individuals will suffice to generate a reliable function.) Guilford[2] discussed the various methods of fractionation and multiplication in his book, *Psychometric Methods*. Other experimental psychology texts written in the 1940s and 1950s also deal with the procedure.[5,6]

B. Absolute Number Matching or Magnitude Estimation

During the later 1940s, S. S. Stevens, then director of Harvard's Psychoacoustic Laboratory, began a series of studies to erect ratio scales using more direct procedures. The fractionation and multiplication methods generated scales, but like Fechner's[7] scaling methods (which aggregated JNDs, or just noticeable differences), the methods were long, tedious, and often frustrating to implement. Researchers needed a simplified method, as easy to use in the laboratory as the conventional fixed-point category scale was in applied field work. In order to realize the potential rewards of simplifying the ratio scaling procedure, one needed only to look at how rapidly researchers expanded their focus and domain once they had the category scale to measure responses.

Stevens experimented with a variety of different scaling methods until he discovered a remarkably simple procedure. He asked panelists simply to match numbers to stimulus intensities so that the ratios of the numbers they assigned matched the perceived or sensory ratios. The approach seemed relatively straightforward, provided that panelists could validly estimate ratios of sensory impressions (e.g., Light ''A'' seems to be three times brighter than Light ''B'', or a similar type of evaluation). Furthermore, Stevens assumed (in 1953, at least) that the panelist's scale of numbers was equivalent in numerical properties to actual numbers (viz., that a 100 represented twice as strong a perception as a 50, so that panelists could validly use numbers to scale perceptions).[8]

The initial experiments revealed that panelists used numbers on a valid ratio scale. The proof was indirect, rather than direct. Stevens based his argument upon the observation that a systematic relation between panelist's ratings and the physical energy level of both tones and lights continued to appear from experiment to experiment (at least based upon the average ratings from panels composed of as few as ten individuals).[9] The relation followed a power law of the form:

$$\text{Perceived intensity} = k(\text{physical intensity})^n$$

In log coordinates, this curved relation becomes a straight line with slope n and intercept log k:

$$\text{Log perceived intensity} = \log k + n(\log \text{physical intensity})$$

According to Stevens, if the exponent n is a constant value for a specific sense continuum (e.g., the brightness of light, the loudness of tones, etc.), then it is improper to assume that the dependent variable (ratings of perceived intensity) lies on an interval scale (with no true 0 point). If the ratings lie in an interval scale, one could take the data from a magnitude estimation study and add a constant to all the ratings. (For an interval scale, this is entirely permissible. For a ratio scale, it is not.) Adding the constant changes the exponent n. Depending upon the constant added, n could drop down to 0 or become extremely high. Yet, as Stevens showed, for each continuum a power function with a reliable exponent n continues to reappear from experiment to experiment when the panelists are permitted to use any numbers that they wish (provided ratios of numbers match ratios of perceived intensities).

The constancy and reliability of the exponent in the power function strongly suggested that the numbers assigned by the panelists possessed ratio properties. The argument by no means swayed all of the parties concerned (including the newly hatched group of vociferous critics) but it did go a long way toward convincing researchers that they might have finally come up with a strong, easily implemented form of measurement leading to the desired ratio scale of perception.

C. Scientific Studies Using the Magnitude Estimation Procedure

Literally thousands of experiments have been done and several hundred technical papers published which use the procedure in one or another of its forms. Marks[10] and Stevens[11] summarized much of the work from the early days (1950s) until the mid 1970s. Since then, the procedure has gained even more acceptance. One is likely to encounter magnitude estimation scaling in scientific journals as diverse as psychology, sociology, medicine, etc. Wherever researchers publish studies on sensory processes or sensory analysis, one is likely to find some investigators using the procedure to uncover the laws of perception.

D. Methodological Studies

The early papers on scaling often focused on the testing procedures and how they affect the resulting ratio scale for perception. One set of studies looked at whether or not panelists had to have a standard stimulus in order to anchor their scale.[12] In these studies, the investigator presented the panelists with a set of stimuli of graded physical magnitude in random order. Some panelists were provided with a first stimulus (the reference or standard) that lay in the middle of the intensity range. Other panelists who evaluated the same set of stimuli were presented with a standard which was at the top of the intensity range. A third group of panelists was presented with a standard that lay at the bottom of the range. As one might expect, the physical intensity of the standard exerted some effect on the ratings. When the standard was the lowest intensity, the panelists used a wide range of numbers, generating a steep slope relating physical magnitudes and sensory ratings. In contrast, when the standard was the stimulus having the highest intensity, the panelists used a smaller range of numbers, generating a flatter curve. The effect of the standard was noticeable, but not overly dramatic.

The standard or first stimulus is important because it sets the reference point for the scale. Other early research focused on whether or not panelists paid attention to the standard.[13] In a set of experiments, the researcher would present the standard first and then the test stimulus. Each pair was composed of the same standard and then a different test stimulus. It soon became obvious that panelists ignored the standard, at least according to their verbal reports after the test session. The panelists concentrated only on the test stimulus. Even though the ratio scale requires a unit of measurement, the panelist does not automatically set his/her scale back to zero each time, and refer anew to the standard. Rather, once the standard and numerical scale become embedded in the panelist's mind, the panelist has little or no trouble assigning ratings. This "automatic" behavior seems much simpler than the agonizing task of comparing each stimulus to the standard, computing the ratio, and then assigning the numbers. The panelist appears to develop a mental scale of magnitudes, easily assessed during the evaluation task.

Other methodological research concerned the relation between this free-number matching and ratio evaluations of pairs of stimuli. One can approach ratio scaling in at least two different fashions. The Stevens approach allowed the panelists to assign numbers freely (subject to the initial instructions). An alternative method (favored by Ekman in Sweden) is to present panelists with pairs of stimuli, and instruct them to estimate the ratio of the sensory magnitudes (between the more intense and the less intense, respectively). Presumably, these two methods should agree. One ought to be able to reconstruct the magnitude estimation scale from the ratio estimation data and vice versa. The scales did agree, confirming that the two sets of instructions and test procedures conditions gave rise to similar scales of perception.[14]

A final methodological study concerns the nature of the instructions. If, as seems to be hinted above, panelists can assign numbers almost "automatically" on a ratio scale, does this imply that panelists possess a built-in scale of magnitude, tapped by the magnitude estimation procedure? Perhaps it is not necessary to instruct panelists to assign numbers in proportion to the ratio of perceptual intensities. Rather, one could simply instruct panelists to match numbers to the magnitudes of their perceptions without ever using the word "ratio". The panelists would automatically use their innate scale, which has the properties of the ratio scale. Indeed, in most of the later work emerging from Harvard's Laboratory of Psychophysics during the later 1960s and early to mid-1970s, the data were derived from panelists using just these simple instructions. The same functional relations between stimulus magnitude and intensity rating emerged, whether the panelists were instructed to assign numbers with ratio properties or were instructed simply to match numbers to perceived intensity. This happy congruence of results led the modern day psychophysicists (under Stevens' continuing encouragment) to propose that they had tapped the basic scale of subjective magnitude and that this scale had clearly defined ratio properties.[15]

E. Developing Functional Relations Between Stimuli and Perceptions

Had magnitude estimation and its allied techniques provided just another method for measurement, it would have excited neither the interest of researchers nor the wrath of critics in the manner it has. However, the discovery that the magnitude estimates generated a repeatable function when plotted against physical measures sparked considerable research. Starting in the early 1950s and continuing in an unabated fashion, researchers used the scaling procedure to uncover the mechanism of the senses. If, in fact, a power function describes the growth of sensory intensity with physical measure, then this "law" or functional relation opens up numerous avenues for research, such as the following:

1. What is the exponent of the power function for each sense modality?
2. Within a single sense modality (e.g., sweetness), do all the relevant stimuli show the same power function parameters, or do they differ?
3. What is the nature of the intensity relation when the panelist evaluates mixtures of stimuli with one stimulus held constant at a level above threshold, while the experimenter systematically varies the other stimulus?
4. Can the procedure be used for other attributes of sensation besides overall intensity? For example, can it be used to measure the degree of pleasantness, rather than strength? Can panelists use the scaling method to partition the overall sensory impression of a complex stimulus into the sense components (e.g., partition a taste impression into the contributions of sweet, salty, sour, and bitter, respectively)?
5. Does the power function remain unchanged when the physiological state of the panelist varies? For instance, when individuals suffer a hearing loss, what are the parameters of their power functions for loudness (vs. sound pressure level) compared to those of normal, nonimpaired panelists? In what ways is the relation the same, and in what ways does it differ? For individuals with kidney failure who undergo salt-restricted diets, does the power law for perceived saltiness (vs. level of sodium chloride) change appreciably?

For researchers, these questions are of great interest. One might be able to answer them satisfactorily with conventional category scales, especially if the scales are broadened to encompass a wide number of allowable categories. (For instance, expanding the traditional 9-point scale to a 100-point scale would do fairly well.) However, it was only with the advent of a ratio scaling technique that the acceleration of sensory research began. Perhaps the growth was due to two factors operating in tandem:

1. A validated ratio scale, the "holy grail" for measurement
2. The discovery of an organizing principle for sensory intensity — namely the power law, the parameters of which were reliable and could be compared across modalities or, within sense modalities, across stimulus types, test conditions, or physiological conditions of the panelists

F. The Power Law as the Organizing Principle

As noted above, when one plots either the geometric mean of the magnitude estimates or the median vs. physical measures, the resulting plot shows curvature. Figure 1 shows some schematic plots. The curves straighten out when one plots the logarithm of the physical measure on the abscissa vs. the logarithm of the average (or median) magnitude estimate on the ordinate.

The results show that sensory intensity tends to increase in a monotonic (viz., continuous upwards) fashion with physical magnitude according to a power function with a defined exponent n and an intercept k (which depends upon the units of measurement of both the stimulus and the panelist's scale).

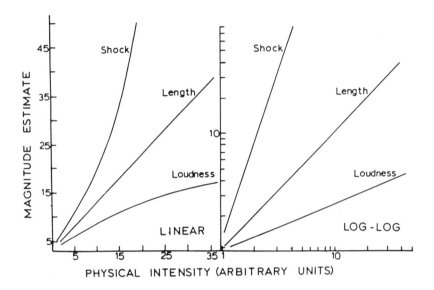

FIGURE 1.   Schematic magnitude estimation function vs. physical intensity. The left panel shows prototypical curves, in which the investigator systematically varied a physical stimulus (e.g., the amperage of an electrical stimulus shock, the length of a line, or the sound pressure level of a tone). The panelists' ratings generate curves. On the right, these curves become straight lines when both the stimulus levels and the mean ratings are converted to their logarithmic equivalents.

Table 3 lists power function exponents for a variety of sensory continua,[16] whereas Table 4 shows the power function exponents for one particular continuum — olfaction — but for a variety of different chemicals presented in various diluents.[17]

Exponents for power functions are assumed (and generally confirmed) to be reliable from study to study and thus provide insight into how we transform physical stimuli into sense magnitude. Although the power function simply describes the transformation and in actuality does not tell us how the sensory system works, we can learn the following:

1.   If the exponent equals 1 (as it does for the perceived length of line vs. actual length), then one responds veridically to the stimulus. When the stimulus changes by 10% in magnitude, the change is perceived to be 10%. The physical range is neither expanded nor contracted when the physical stimulus is transformed to a perception.

2.   If the exponent is less than 1 (as it is for the perceived odor intensity of virtually all chemicals, the perceived viscosity of liquids, etc.) then the physical range of magnitudes is compressed into a smaller sensory range. For example, suppose the exponent is 0.4 (as it is approximately for the perceived viscosity of silicone oils). When two stimuli lying in a ratio of 1000 to 100 centipoises (CP) (a measure of viscosity) are compared, these stimuli are not perceived to lie in a ratio of 10:1. The more viscous stimulus seems *far less* than ten times thicker. In fact, it seems only $(1000/100)^{0.4}$, or only about 2.5 times as viscous. The large ratio that is measured by objective means transforms to a smaller perceptual ratio.

3.   If the exponent exceeds 1 (as it does for the perceived roughness vs. the grit value of sandpaper), then the physical range is expanded to a wide range of perceptual intensity. For electric shock, the exponent has been measured to be 2.5, or more. Small changes in grit or in the electrical current produced expanded impressions and a larger ratio of perceived magnitude.

Table 5 shows how exponents govern the change in perceived intensity.

TABLE 3
REPRESENTATIVE EXPONENTS FOR POWER FUNCTIONS
RELATING SENSORY MAGNITUDE TO STIMULUS INTENSITY[16]

Continuum	Measured exponent	Stimulus condition
Loudness	0.67	Sound pressure of 3000-Hz tone
Vibration	0.95	Amplitude of 60 Hz on finger
	0.6	Amplitude of 250 Hz on finger
Brightness	0.33	5° target in dark
	0.5	Point source
	0.5	Brief flash
	1.0	Point source briefly flashed
Lightness	1.2	Reflectance of gray papers
Visual length	1.0	Projected line
Visual area	0.7	Projected square
Redness (saturation)	1.7	Red-gray mixture
Taste	1.3	Sucrose
	1.4	Salt
	0.8	Saccharine
Smell	0.6	Heptane
Cold	1.0	Metal contact on arm
Warmth	1.6	Metal contact on arm
	1.3	Irradiation of skin, small area
	0.7	Irradiation of skin, large area
Discomfort, cold	1.7	Whole body irradiation
Discomfort, warm	0.7	Whole body irradiation
Thermal pain	1.0	Radiant heat on skin
Tactual roughness	1.5	Rubbing emery cloths
Tactual hardness	0.8	Squeezing rubber
Finger span	1.3	Thickness of blocks
Pressure on palm	1.1	Static force on skin
Muscle force	1.7	Static contractions
Heaviness	1.45	Lifted weights
Viscosity	0.42	Stirring silicone fluids
Electric shock	3.5	Current through fingers
Vocal effort	1.1	Vocal sound pressure
Angular acceleration	1.4	5-sec rotation
Duration	1.1	White noise stimuli

II. A MAGNITUDE ESTIMATION EXPERIMENT OF PERCEIVED ODOR INTENSITY OF PROPANOL

This section illustrates how psychophysicists use magnitude estimation to assess the relation between apparent physical viscosity and perceived sensory viscosity. Although the section deals only with one sensory continuum, the odor strength of *n*-propyl alcohol (propanol), the approach is similar to the many hundreds of studies published in the scientific literature.

A. Selection of Stimuli

The researcher using magnitude estimation investigates the perception of supraliminal (above-threshold) levels of odor intensity produced by propanol at varying percent saturation. In contrast to threshold studies, the researcher uses stimuli that are noticeably above threshold — which in this case means stimuli which have a distinct and perceptible smell. For studies of odor and taste perception, the investigator should make sure that the panelist can actually sense the stimulus before performing the actual experiment.

Researchers try to use as wide a range of physical stimuli as possible and practical. Rather

Table 4
EXPONENTS (N) FOR THE
POWER FUNCTION RELATING
PERCEIVED ODOR INTENSITY
(S) TO PHYSICAL ORORANT
CONCENTRATION (C)[17]

Odorant	Exponent	Diluent
Amyl acetate	0.13	Liquid
Anethole	0.16	Liquid
Butanol	0.31	Liquid
	0 .64	Air
	0.66	Air
Butyl acetate	0.58	Air
Butyric acid	0.22	Liquid
Coumarin	0.33	Air
Citral	0.17	Liquid
Ethyl acetate	0.21	Liquid
Eugenol	0.27	Liquid
	0.64	Air
Geraniol	0.20	Air
Guaiacol	0.20	Liquid
Heptanol	0.16	Liquid
Hexanol	0.15	Air
D-Menthol	0.24	Liquid
Methyl salicylate	0.20	Liquid
Octanol	0.24	Liquid
Pentanol	0.21	Liquid
Phenylethanol	0.19	Liquid
Phenylacetic acid	0.12	Liquid
Propanol	0.52	Air
Isovaleric acid	0.21	Liquid

than constraining the stimuli to lie within a "narrow window" of the entire range, the researcher develops a sensory function that categorizes a wide range. Quite often the function changes in its curvature at the very lowest or highest levels.

For this particular study, the researcher chose to test six odor stimuli, varying from 1.7% saturation of *n*-propanol in air to 40% saturation. The levels were 1.7, 3.6, 6.6, 12.3, 21.7, and 40%. Note that the stimuli are approximately equally spaced in terms of ratios, not intervals. Since the relation between sensory magnitudes and physical magnitudes approximates a power law rather than a linear function, equal physical ratios generate equal sensory ratios. In log-log coordinates, where the function straightens into a line, these equal physical ratios become equal distances on the scale.

B. Instructions

Over the past 30 years, researchers have used a variety of instructions for magnitude estimation, ranging from detailed instructions about using ratios of numbers down to the simplistic request to "match numbers to perceived intensity" (without ever mentioning the word ratio). Most of these instructions provided essentially the same results. As a consequence, the instructions shown in Table 6 suffice.

C. Orientation in Scaling

Sometimes researchers feel that panelists need to be oriented in evaluation by means of a short warm-up exercise. For instance, one can orient panelists by showing them circles of varying area and asking them to match numbers to the circles so that the larger the circle

Table 5
HOW THE POWER FUNCTION EXPONENT GOVERNS THE TRANSFORMATION OF PHYSICAL TO PERCEPTUAL RATIOS

	Physical ratio between stimuli						
Exponent	**2.00**	**4.00**	**6.00**	**8.00**	**10.00**	**20.00**	**30.00**
−2.00	0.25	0.06	0.03	0.02	0.01	0.00	0.00
−1.80	0.29	0.08	0.04	0.02	0.02	0.00	0.00
−1.60	0.33	0.11	0.06	0.04	0.03	0.01	0.00
−1.40	0.38	0.14	0.08	0.05	0.04	0.02	0.01
−1.20	0.44	0.19	0.12	0.08	0.06	0.03	0.02
−1.00	0.50	0.25	0.17	0.13	0.10	0.05	0.03
−0.80	0.57	0.33	0.24	0.19	0.16	0.09	0.07
−0.60	0.66	0.44	0.34	0.29	0.25	0.17	0.13
−0.40	0.76	0.57	0.49	0.44	0.40	0.30	0.26
−0.20	0.87	0.76	0.70	0.66	0.63	0.55	0.51
0.00	1.00	1.00	1.00	1.00	1.00	1.00	1.00
0.20	1.15	1.32	1.43	1.52	1.58	1.82	1.97
0.40	1.32	1.74	2.05	2.30	2.51	3.31	3.90
0.60	1.52	2.30	2.93	3.48	3.98	6.03	7.70
0.80	1.74	3.03	4.19	5.28	6.31	10.99	15.19
1.00	2.00	4.00	6.00	8.00	10.00	20.00	30.00
1.20	2.30	5.28	8.59	12.13	15.85	36.41	59.23
1.40	2.64	6.96	12.29	18.38	25.12	66.29	117
1.60	3.03	9.19	17.58	27.86	39.81	121	231
1.80	3.48	12.13	25.16	42.22	63.10	220	456
2.00	4.00	16.00	36.00	64.00	100	400	900

Note: Numbers in the body of the table are perceptual ratios corresponding to the following quantity: (physical ratio)^{exponent}.

Table 6
INSTRUCTIONS[a]

You will be evaluating a set of odorants which vary in intensity (strength). There are a total of 12 odorants.[a] Please evaluate each odorant in the order shown on the sequence sheet which accompanies this set of instructions.

The Scale
You may use any positive numbers that you wish. Match numbers to the strength of the odor intensity as you personally perceive it. You may go as high as you want to show increasing strengths of odor intensity. Zero (0) on the scale means that you do not perceive any odor at all — that is, there is no smell at all.

Please match numbers to odor intensity so that ratios of your ratings match ratios of the strengths of the odor as you perceive the odor to be.

[a]   Each panelist scaled the six odorant levels of *n*-propanol two times in a randomized order.

is, the higher the number will be. This orientation does not create "experts" out of the panelists, but rather eliminates any potential source of confusion that may exist in the panelist's mind. By inspecting the ratings assigned by panelists during the warm-up exercise, one can rapidly determine whether or not the particular panelist has grasped the concept of magnitude. (Small circles should be assigned small numbers, and large circles should be assigned large numbers.)

D. Presentation of the Stimuli

The traditional and easiest way to present stimuli consists of presenting the stimuli singly

Table 7
NORMALIZATION PROCEDURES FOR MAGNITUDE ESTIMATION RATINGS

Method 1 — Modulus Equalization

Objective

Normalize all the panelist's ratings, so that the geometric mean of each panelist equals the geometric mean of every other panelist. All geometric means equal a predefined constant.

Requirements

All of the panelists must each evaluate the same stimuli.

Approach

For each panelist, compute the product of the panelist's ratings on the set of stimuli.

Compute the logarithm of the product (obtained in step 1 above), and compute the mean. This is the logarithm of the geometric mean.

Decide the final geometric mean that each panelist will have. Call this log geometric mean "X".

For each panelist compute the following correction factor:

$F = \overline{X} - \log$ (target geometric mean)

where $\overline{X}$ is the log geometric mean for the panelist. Add this correction factor to every one of the panelist's ratings. Compute the antilogarithm of the ratings. (Note that computationally, one needs only the geometric mean from each panelist and the target geometric mean).

Method 2 — Modulus Normalization

Assumption

Each panelist tests one stimulus in common with every other panelist. However, the remaining stimuli may differ from panelist to panelist.

Approach

For each panelist, locate the common or pivot stimulus. Call this value Y.

Decide upon a common target value for Y.

For a specific panelist who assigned the common stimulus a value of Z, compute a correction factor as follows:

Correction factor = Y/Z

Multiply all of that panelist's rating by the correction factor.

Develop a unique correction factor for each panelist and follow the two preceding steps for the panelist's data.

in a randomized order. It is generally best to begin with a stimulus lying somewhere in the middle range of magnitude, although with a sufficient number of panelists, each stimulus (including the extremes) can be the first stimulus (or standard). Furthermore, each panelist should receive the stimuli in a unique randomized order, so that every one of the six stimuli has a chance to appear equally often in every position. It is not critical to balance the order of appearance precisely, since order effects are marginal (except, perhaps, for the first stimulus because panelists as yet have no frame of reference for their ratings). As long as the order is randomized and reasonably balanced, the investigator has adequately reduced the possibility of bias due to presentation sequence.

E. Data Normalization — Bringing the Ratings to a Common Scale Across Panelists

In traditional psychophysical studies, panelists rate each of the stimuli at least one time. They follow a ''complete block'' design in the statistical sense. In these experiments there is no need to present each panelist with an incomplete set of the products. Either normalization procedure shown in Table 7 (Method 1 or Method 2) will work. Since panelists use different scales and different numbers to rate intensity, substantial interpanelist variability exists in the data, traceable to the different sizes of the numbers. Normalization reduces this unwanted variability by making the size of numbers similar across panelists. It does so by an allowable multiplicative transformation.

When each panelist rates an *incomplete set* of the stimuli, it is necessary to include a common stimulus in every experiment for every panelist. (This may be one or several stimuli common to all panelists and experiments in the study.) The common stimulus need not be identified. One would than "normalize" all data in accordance with the approach shown in Table 7 (Method 2). Note that the normalization shown in Table 7 (Method 2) works equally well when every panelist rates all of the stimuli. It is still possible to normalize the data on the basis of one or a limited number of stimuli chosen as implicit "references" within the stimulus set tested.

Table 8 shows the ratings from ten panelists for perceived odor intensity assigned to the different stimuli. Panel A shows the raw data and the geometric mean. Panel B shows the data converted to logarithms. Panel C shows the equalization procedure, with the data (in logarithms) converted by Method 1. Finally, Panel D shows the linear equivalents of the logarithmic values in Panel C. Note that the ratios among numbers for each panelist remain the same, but the sizes of the numbers become similar.

The geometric mean is the appropriate measure of central tendency because a scatter plot of the ratings shows that the magnitude estimates are distributed "log normally". That is, in logarithmic paper, plotting the ratings reveals a normal distribution. When ratings are distributed "log normally", the geometric mean is the appropriate measure. If the geometric mean is not appropriate (e.g., if there are 0 ratings because some panelists could not detect the relevant intensity), then the next best statistic is the median.

Typically, the final analysis if the data consists of graphing the relation between attribute rating and physical magnitude or computing the slope by regression analysis. Of key interest from the regression analysis is the slope of the function, which turns out to be 0.66. This means that perceived odor intensity of *n*-propanol grows with physical concentration measured objectively, but at a slower rate. Table 5 shows the expected change in perceived magnitude for various sensory continua governed by exponents ranging from 0.2 to 2.0. With an exponent of 0.66, doubling the physical level generates only a 1.58-fold increase in perceived magnitude.

F. Other Types of Analysis and Experimental Designs

The foregoing experiment concerned the perception of a single sensory continua — perceived odor intensity — as a function of concentration along a single phsyical continuum — percent saturation in air. There are other designs and questions that this experimental paradigm can answer. Some of them are

1. What is the relation between an attribute, its inverse, and physical magnitude? Suppose the panelists scaled the attribute of "weakness of odor" rather than "intensity". Would the psychophysical function have been the reciprocal (showing up as a function with exponent −n rather than +n)? Quite often, researchers who investigate attributes and their inverses find that the inverse attribute (e.g., softness vs. loudness) is not precisely an inverse when it comes to the function relating sensory attribute level and physical magnitude. Thinness or wateriness instead of viscosity, or softness instead of loudness decreases with increasing physical magnitude (viz., apparent physical viscosity or sound pressure level). However, the function plotted in log-log coordinates (which straightens out curvature for the "regular" attribute) shows curvature near the bottom for the "inverse" attribute. It is as if at very low stimulus levels panelists lose their ability to accurately rate thinness (vs. physical viscosity), softness (vs. sound pressure), or smallness (vs. physical area). Researchers do not know why the inverse attribute exhibits the distortion, whereas the "regular" attribute exhibits much more linearity.[18]

2. What are the intensity functions for two different types of stimuli generating the same sensory response? Table 4 shows that for olfactory perception, different chemicals generate a diversity of exponents. Even though all chemicals excite olfactory sensations, they follow different "laws". Similarly, for the sense of taste, different sweeteners will generate a diversity of exponents when all are dissolved in water and tested

Table 8
DATA MATRIX FOR MAGNITUDE ESTIMATION RATINGS OF *N*-PROPANOL BY TEN CONSUMER PANELISTS

Panel A — Raw Data Matrix

			Stimulus levels				
Panelist	A 1.7	B 3.6	C 6.6	D 12.3	E 21.7	F 40	Geometric mean
1	30	15	22	30	50	60	31.0
2	10	8	10	23	24	80	18.7
3	1	2	18	12	18	25	7.6
4	10	15	20	15	25	30	18.0
5	3	15	20	55	43	65	22.7
6	4	5	15	18	30	33	13.2
7	20	10	50	55	160	200	51.0
8	4	2	12	10	75	100	13.9
9	5	5	15	25	25	50	15.1
10	2	3	20	10	35	45	11.1
Geometric mean	5.57	6.18	18.33	21.03	38.58	56.77	

Panel B — Logarithmic Values of Magnitude Estimates (Prenormalization)

Panelists	A	B	C	D	E	F	Correction factor
1	1.48	1.18	1.34	1.48	1.70	1.78	−0.49
2	1.00	0.90	1.00	1.36	1.38	1.90	−0.26
3	0.00	0.30	1.26	1.08	1.26	1.40	+0.12
4	1.00	1.18	1.30	1.18	1.40	1.48	−0.25
5	0.48	1.18	1.30	1.74	1.63	1.81	−0.36
6	0.60	0.70	1.18	1.26	1.48	1.52	−0.12
7	1.30	1.00	1.70	1.74	2.20	2.30	−0.71
8	0.60	0.30	1.08	1.00	1.88	2.00	−0.14
9	0.70	0.70	1.18	1.40	1.40	1.70	−0.18
10	0.30	0.48	1.30	1.00	1.54	1.65	−0.05

Panel C — Logarithmic Values of Magnitude Estimates (Postnormalization)

Panelist	A	B	C	D	E	F	Total
1	0.99	0.68	0.85	0.99	1.21	1.29	6.00
2	0.74	0.65	0.74	1.10	1.12	1.71	6.00
3	0.12	0.42	1.37	1.20	1.37	1.52	6.00
4	0.74	0.92	1.04	0.92	1.14	1.22	6.00
5	0.12	0.82	0.94	1.38	1.28	1.46	6.00
6	0.48	0.58	1.05	1.13	1.36	1.40	6.00
7	0.59	0.29	0.99	1.03	1.50	1.59	6.00
8	0.46	0.16	0.94	0.86	1.73	1.86	6.00
9	0.52	0.52	1.00	1.22	1.22	1.52	6.00
10	0.25	0.43	1.25	0.95	1.50	1.61	6.00

Table 8 (continued)
DATA MATRIX FOR MAGNITUDE ESTIMATION RATINGS
OF *N*-PROPANOL BY TEN CONSUMER PANELISTS
Panel D — Magnitude Estimates (Postnormalization)[a]

Panelist	A	B	C	D	E	F
1	10	5	7	10	16	19
2	5	4	5	12	13	44
3	1	3	24	16	24	33
4	6	8	11	8	14	17
5	1	7	9	24	19	28
6	3	4	11	14	23	25
7	4	2	10	11	31	39
8	3	1	8	7	52	69
9	3	3	10	17	17	33
10	2	3	18	9	31	40
Geometric mean	3.3	4.7	10.0	13.1	17.6	26.7

Note: Each data point is the geometric mean of a panelist who rated the stimuli twice. The geometric mean for each panelist equals 10.

[a] To nearest integer value.

in the same ratios (e.g., a concentration series of five stimuli, each stimulus being two times more concentrated than the one below it).[19] The diversity of exponents governs the chemical senses, depending on the chemical. For other senses, such as the tactile perception of thickness or the loudness of noises the exponent is more constant across different types of stimuli. For studies of loudness, the exponent tends to be similar whether the investigator studies pure tones, white noise, or a mixture of tones. As yet, researchers do not know the reasons for the divergence of exponents for the same sensations in taste and smell due to variations in the chemical stimulus. We can speculate that the physical measure of stimulus concentration, equally applicable to all chemicals (e.g., percent saturation, molarity, parts per million) is really not the effective stimulus that excites the receptor. Thus, whereas the researcher might investigate the perception of sweetness of both sugar and aspartame, the concentration measures in the liquid may have relatively little to do with the effective concentration of these sweet agents on the tongue.

III. MAGNITUDE ESTIMATION OF LIKING

Although the early work on magnitude estimation concentrated upon sensory intensity functions, it was just over 10 years afterwards that researchers became interested in using the scaling method for measurements of liking. Some early studies on hedonics involved tastants[20] and odorants.[21] As was the case with intensity functions, the researchers investigated "model systems" (pure chemicals) rather than real foods or fragrances. They searched for the relation between degree of liking and either stimulus intensity or perceived sensory intensity, respectively.

A. The Inverted U Curve as a General Model for Liking Functions
Studies using magnitude estimation, both of sensory intensity and of liking, revealed that there exists surprising lawfulness in hedonics. As a stimulus increases in physical or perceived magnitude, liking goes up, peaks, and then goes down. This inverted, U-shaped curve,

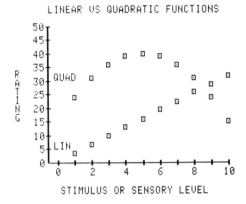

FIGURE 2.   Schematic functions for perceived intensity and degree of liking. Panel ratings for both
attributes appear on the ordinate, vs. the physical
concentration of the stimulus (abscissa). The specific
parameters of the functions vary with the actual
physical stimulus.

shown schematically in Figure 2, does not depend upon the magnitude estimation of scaling
to appear. Category scales would do just as well. Indeed, polling the population to determine
the percentage of likers of various sugar solutions at different concentrations shows the same
type of inverted U pattern.[22] Nonetheless, it was the psychophysical scaling method of
magnitude estimation which accentuated this recurring lawfulness and tried to fit an equation
to the data. Other researchers noted the inverted U-shaped curve, but did not look for the
parameters of the curve.

B. Unipolar vs. Bipolar Scales to Measure Liking

A critical problem in hedonics is the opposition to liking and disliking as attributes. Is
disliking simply the total absence of liking (equivalent, therefore, to a 0 on the magnitude
estimation scale)? Or do liking and disliking represent two opposite continua, each with its
own rules?[23]

At first glance, the foregoing issue appears to be only a semantic question. Does it really
matter whether the hedonic scale comprises graded degrees of liking, with disliking defined
to be 0? For intensity scaling we know that a stimulus having no intensity is rated as 0. The
researcher can adjust the range of stimuli to ensure that all tested stimuli have a perceived
intensity (viz., all stimuli lie above threshold). The problem becomes more difficult in
hedonics. The same stimulus can be very pleasant to one individual and very unpleasant for
another. If unpleasantness, at any level, is assigned a 0, then no matter what degree of
unpleasantness is experienced, the panelist must assign that stimulus a 0. The scale captures
lack of any pleasantness and increasing (graded) degrees of pleasantness. If, on the other
hand, pleasantness and unpleasantness each have magnitudes, then what scale can encompass
both? The traditional (unipolar) magnitude estimation scale with an origin at 0 cannot,
because the panelist must divide the scale into two parts (corresponding to the section for
liking or pleasantness and the section for disliking or unpleasantness). Zero is not appropriate
at the demarcation point, nor is any positive nonzero number appropriate to divide the region
of liking or disliking.

The foregoing problems, contrasting unipolar vs. bipolar liking scales, gave rise to two
different methods for scaling degree of acceptance. In the first (unipolar) method, panelists
simply rate the degree of liking, assigning 0 to low levels of liking and high levels to high
degrees of liking. In the second (bipolar) method, panelists had to make two assignments

Table 9A
INSTRUCTIONS FOR UNIPOLAR MAGNITUDE ESTIMATION OF LIKING

Please assign numbers to reflect how much you like or dislike the stimuli. Your scale starts at 0, which denotes disliking (no liking at all). As your level of liking increases, your numbers should increase as well. If you rate one stimulus 120, and another 60, then you like the first stimulus twice as much as you like the second.
Be sure to use only numbers from 0 upwards, with 0 denoting no liking at all.

Table 9B
INSTRUCTIONS FOR BIPOLAR MAGNITUDE ESTIMATION OF LIKING

You will be evaluating a number of samples. We would like you to rate how much you like or dislike each sample using the following scale.
First, in the box marked D or L, write down the letter L if you like the sample, or write down the letter D if you dislike the sample. If you feel neutral — that is, you neither like nor dislike the sample — then write down a 0 in that box.
Now, where it says "amount", think about how much you like or dislike the sample. If you like it or dislike it just a little, write down a small number. If you like it or dislike it a lot, write down a big number.
The examples below show you what is meant.

Dislike a lot	L/D	☐	and a Big number	☐
Dislike a little	L/D	☐	and a Small number	☐
Neutral	L/D	☐	and the number "0"	☐
Like a little	L/D	☐	and a Small number	☐
Like a lot	L/D	☐	and a Big number	☐

in their rating task. First, they had to decide whether or not they liked the stimulus. This decision was coded as a + (or the letter L) for liking, or − (or the letter D) for disliking. Tables 9A and 9B contrast the two approaches in terms of the instructions to the panelists.

The two methods for scaling degree of liking generate similarly appearing patterns for liking vs. the magnitude of the stimulus. Whether the panelist uses only positive numbers or uses a bipolar scale, the relation follows an inverted, U-shaped curve. The parameters of that curve (e.g., the steepness) vary with the scaling method. For sweeteners, however, the concentration at which liking reaches its maximum level did not change, even though the panelists used two different types of scales.[23-25]

C. Representative Hedonic Curves Using Magnitude Estimation

Figures 3 and 4 show typical magnitude estimation curves for the liking of various stimulus vs. physical magnitude.[26] Figure 5 shows how panelists who are chronically restricted in salt intake due to kidney failure scale degree of liking for salt added to a salt-free soup.[27] The control group of panelists, without salt-restricted diets, shows a different curve for liking vs. concentration.

IV. OUT OF THE LABORATORY — INTO THE WORLD OF PRODUCT TESTING

A. Background

The previous studies and précis of magnitude estimation cover the period up to the mid-1970s. Until then, magnitude estimation had remained a scaling technique primarily within the purview of a limited number of academically oriented researchers. One had to consult archival periodicals in the scientific disciplines in order to see uses of the scaling method and its accomplishments.

As scientists become familiar with their tools and achieve success with them, they broaden their perspectives. As the 1970s progressed, many researchers in food companies became interested in better ways to measure sensory reactions to foods. Most practitioners used

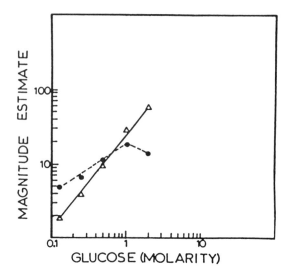

FIGURE 3.   Relation between the molarity of glucose and the perceived sweetness (open triangles) and the degree of liking (filled circles). The data is averaged from several hundred panelists over a wide number of studies.[25]

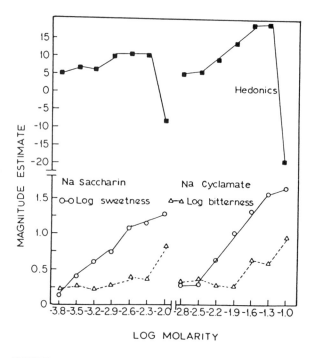

FIGURE 4.   Relation between the log molarity of two artificial sweeteners (Na saccharin; Na cyclamate), ratings of sweetness and bitterness (bottom), and degree of liking (positive numbers) and disliking (negative numbers). Sweetness, bitterness, and liking were all obtained using the magnitude estimation scale. The sweetness and bitterness ratings are plotted in logarithmic values vs. the log molarity. The liking ratings themselves are plotted vs. the log molarity.[26]

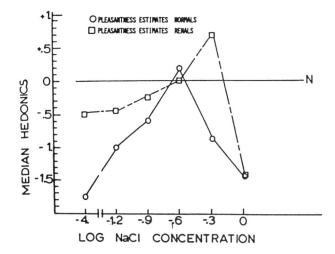

FIGURE 5.   Relation between the log NaCl concentration is soup and degree of liking. Liking was rated on a bipolar magnitude estimation scale and normalized to a scale of -2 (dislike extremely) to 0 (neutral) to $+2$ (like extremely). The data compare the ratings assigned by two groups of panelists. One group consisted of normal volunteers. The second group consisted of panelists with renal failure who were undergoing dialysis treatments.[27]

either a limited category scale (e.g., 1 to 5 or 1 to 9) or a linear scale. Researchers were ready for a better method of measuring magnitudes of perception.

The mid-1970s saw the emergence of magnitude estimation as a tool for industry. Researchers had to overcome several key problems:

1.   Whereas scientific researchers work with well-defined physical stimuli, applied researchers work with complex stimuli which often cannot be systematically varied. Would the magnitude estimation method prove as useful for these complex stimuli, or would it turn out to be limited to model systems, varying only on one physical dimension?

2.   Whereas scientific researchers instructed panelists to scale the intensity of only one attribute (or at most two, e.g., intensity and liking), applied researchers want panelists to evaluate a large set of attributes for a single product (or at most a small array of products). Could panelists validly rate a product on a battery of attributes and obey the ratio requirements of the scale? These attributes tap sensory and hedonic responses for appearance, aroma, taste, and texture.

3.   Psychophysicists have the luxury of testing all the stimuli with a panelist in a single session. Each panelist evaluates every stimulus. However, in applied research this "complete-block" design often is not feasible. A panelist might test only half, a third, or even a smaller fraction of the stimuli in the typical applied setting (e.g., at a company plant). Given the fact that each panelist could generate his/her own scale and the fact that each product would be tested by a different subset of panelists, how could the investigator be sure that the differences in ratings between two products stemmed from real product differences, and not from differences in scales chosen by different panelists? Magnitude estimation allows each panelist to assign numbers using his/her own scale. The interpanelist variability due to scale usage may exceed the interproduct variability.

Table 10
COMPARISON OF MEAN
RATINGS FOR COMPLETE
VS. INCOMPLETE BLOCK
DESIGN

Full Block[a]

Panelist	Samples		
	A	B	C
1	1	2	3
2	10	20	30
3	40	80	120
4	100	200	300
5	200	400	600
Means			
Arithmetic	70	140	210
Geometric	24	48	72
Median	40	80	120

Incomplete Block[b]

Panelist	A	B	C
1	—	2	3
2	10	—	30
3	40	80	—
4	—	200	300
5	200	—	600
Means			
Arithmetic	83	94	233
Geometric	41	31	63
Median	40	80	165

Note: Numbers in body of table are magnitude estimates.

[a] All panelists rate A, B, and C.
[b] Each panelist rates two of three.

Table 10 compares the results for three stimuli, A, B, and C, rated in a complete block vs. incomplete block design. Note the real possibility that without an adequate normalizing procedure, two products might generate misleading results.

B. Normalizing Magnitude Estimates by External Verbal Standards
One can address the issue of incomplete designs by normalizing the individual panelist's magnitude estimates with an external standard. In the study, each of the panelists uses his/her own scale with a representative and perhaps nonoverlapping group of products (or even with one product). Some type of *external standard* is required by which the scales of different individuals are brought together. This standard must be identical for all individuals. Furthermore, it must be *independent* of the stimuli tested since each individual evaluates a different subset of the stimuli. (Keep in mind that with 20 or more stimuli to test, each individual might evaluate as few as 1 to 5. Some individuals would rate entirely nonoverlapping sets of stimuli.)
The answer to the problem consists of providing each individual with a single word or

Table 11
INSTRUCTIONS FOR CALIBRATING MAGNITUDE ESTIMATES

Amount scale
Looking back at all the stimuli that you have tested today, please assign a number from your scale for a product that would be:

Extremely strong	_____
Very strong	_____
Moderate	_____
Weak	_____
Cannot perceive (call this 0)	0

Your numbers should be on the same scale that you used to rate the samples you tried today. Furthermore, the rating for "Extremely strong" should be the top of your personal scale. No sample can be higher than "Extremely strong".

Liking scale
Looking back at all the stimuli that you have tested today, please assign a number from your liking scale for a product that you feel you:

Like extremely (the top of you liking scale)	+
Like very much	+
Like moderately	+
Like slightly	+
Neither like nor dislike (this is defined as 0 — write in a 0)	0
Dislike slightly	−
Dislike moderately	−
Dislike very much	−
Dislike Extremely	−

set of words at the end of the evaluation. These words denote gradations of intensity. For each word, the panelist rates the word as though it represented yet another stimulus. Thus, if the panelist rates flavorings of varying intensities, then at the end of the evaluation the researcher presents the panelists with a series of words and asks the panelist to rate these words as if they represented flavorings of various intensities. Table 11 shows how the question is asked.

At this point each panelist has provided ratings for a set of samples and has rated the same word or words as if they were additional stimulus samples. One can normalize the individual panelist's verbal scales of magnitude, using the procedure shown in Table 12. Each panelist's ratings are multiplied by a constant (unique for that panelist), so that the top of each panelist's verbal scale is 100. This same multiplier or normalizing constant is then used to adjust that panelist's ratings which were assigned to the actual products, thus bringing them into line (column B).

Researchers have proposed a number of ways of normalizing the data, whether this be by averaging a whole series of different words together (Table 12, column C) or by simply using the panelist's rating for "top of scale" (Table 12, column B). There is no single best way for normalizing. On the other hand, each of the methods will bring into congruence the scales assigned by different panelists, even if those panelists never evaluated the same products.

C. Averaging the Magnitude Estimates

The appropriate averaging procedure for magnitude estimation is the geometric mean, since the magnitude estimates are log normally distributed. The median is second best. Normalized magnitude estimates generate the same log-normal distribution, but the dispersion is much lower than would be the case were the data not normalized. For most applied

Table 12
EXAMPLES OF CALIBRATION

Intensity or Unipolar Liking

	A Precalibration ratings	B Postcalibration word "extreme"[a]	C Postcali- bration all words[b]
Product			
A	36	16	31
B	75	34	64
C	122	55	105
Calibration words			
Extreme	220	100	189
Very	140	63	120
Moderate	70	31	60
Slight	36	16	31
None	0	0	0
Index number		220	116

Bipolar Liking

	A	B	C
Product			
A	40	22	38
B	−120	−66	−114
C	26	14	24
Calibration words			
Like extremely	180	100	172
Like very much	120	66	114
Like moderately	60	33	57
Like slightly	40	22	38
Neutral	0	0	0
Dislike extremely	−55	−30	−52
Dislike Moderately	−80	−44	−76
Dislike very much	−130	−72	−124
Dislike extremely	−170	−94	−162
Index number		180	104

[a] Normalized magnitude estimate $= \dfrac{\text{Pecalibration}}{\text{Index number}}$ X 100. Index number = "Extreme".

[b] Index number = (Extreme + Very + Moderate + Slight) /4. For intensity
Index number = Absolute values of (Extreme + Very + Moderate + Slight) /8 (Using both liking and disliking parts of the calibration scale).

research, the investigators neither use the geometric mean nor make logarithmic transformations. Furthermore, when dealing with negative numbers, the geometric mean and the logarithmic transformations are not appropriate. Most researchers using magnitude estimation for product testing opt to use the conventional statistics. They compute arithmetic means along with statistical tests predicated upon an underlying normal distribution. As long as the data are normalized, the geometric mean and the arithmetic mean are reasonably close to each other, so that using the arithmetic mean does not unduly compromise the data (see Table 13).

Table 13
COMPARISON OF GEOMETRIC VS.
ARITHMETIC MEANS[a]

Panelist	Stimulus					
	A	B	C	D	E	F
1	40	49	51	63	56	55
2	39	41	35	62	44	39
3	37	42	58	68	57	54
4	39	45	40	75	44	36
5	34	43	37	33	40	45
6	39	46	19	29	21	20
7	32	37	40	29	30	32
8	38	37	22	31	22	17
9	50	56	61	70	58	58
10	41	47	43	37	41	43
11	41	42	47	54	53	53
12	36	37	48	60	45	41
13	40	46	50	56	54	53
14	38	45	37	51	42	37
15	41	49	54	62	55	54
Geometric mean	39	44	41	49	42	40
Arithmetic mean	39	44	43	52	44	42

[a]   Normalized data.

V. USING THE MAGNITUDE ESTIMATION PROCEDURE FOR A PRODUCT TEST

A. Background

The study reported below shows how magnitude estimation has been used in product test. The study involves the evaluation of different margarines currently in the marketplace, along with prototypes. In the U.S. market, there are a number of margarine manufacturers who promote many competitive brands. The purpose of this study was to compare the attributes of the various margarines and where relevant to show how each product can be improved.

B. Stimuli

The stimuli included eight different margarine samples tested "blind" (viz., without identification). Prototypes B and E were test products. Six of the products in the test were well-known brands in the market.

All margarines were tested on unsalted crackers. The blind-test products were identified only by a three-digit code number accompanying the product. However, to simplify discussion and data presentation, the samples are listed here by an identifying letter.

C. Panel Composition and Activities

In commercial product evaluation, where interest is focused on actual in-market products or prototypes, the composition of the panel is critical. Investigators typically choose as panelists only those individuals who purchase and/or consume the product. The data lose their validity (in terms of representing reactions of the "true" consumer) if the investigator settles for a "convenience sample" of available participants who are not necessarily "category users".

A total of 200 panelists participated. All were margarine users. They were further classified by age, market, income, and the brand that they used most often. Quite often in studies

Table 14
ACTIVITIES DURING THE MARGARINE EVALUATION STUDY

Panelists were prerecruited by telephone. It they qualified (viz., they used margarines), they were invited to participate by the telephone interviewer.

Panelists showed up for a 4-hr test session at a local testing facility (in a mall).

An attending interviewer oriented the panelists in magnitude estimation by means of two practice exercises:
1. Evaluation of the perceived areas of circles
2. Evaluation of the liking or disliking of actors on television

Panelist ratings were checked for completeness and the panelists were asked a question about their ratings. The questioning process ensured that panelists understood the scale and the activities. Neither the panelist nor the interviewer knew the "correct" answer to the question, but the questioning maintained alertness and motivation.

The panelist rated the first of eight margarines on a "blind" basis in randomized order (unique order for each panelist). The panelist rated the margarine on the attributes.

The ratings were checked for completeness, and the respondent was asked a question about the ratings.

The panelist continued, following the previous format, until she had rated all eight products on a blind basis.

The panelist profiled the "ideal" margarine on the same attributes.

The panelist rated the relative importance of attributes.

The panelist filled out an attitude and usage questionnaire.

The panelist was paid and dismissed.

with commercial products, it is important to specify the sample, so that if one later wishes to look at breakouts of the population into various groups it is possible to do so. For instance, one may wish to look at older vs. younger consumers, or heavy vs. light users of frankfurters. In most scientific studies concerned with "perceptual laws", rather than product acceptance, one rarely looks at individuals or key subgroups in the population. (Note that this difference between basic research and applied research analyses holds for any type of scaling, not just magnitude estimation.)

D. Activities During the Evaluation

Magnitude estimation studies involving actual food products follow similar protocols as do magnitude estimation studies involving model systems. The test session also included ratings of the "ideal product" and the relative importance of various attributes for acceptance. Table 14 shows the activities during the session. Table 15 shows the questionnaire.

E. Analysis of the Ratings

Since each panelist rated all of the products on an open-ended scale, it was necessary to normalize the data. The normalizing of index number is a function of all ratings in the calibration scale. (See Table 12, column C for the actual normalizing method used here.)

Table 16 shows the data base of ratings after normalization. With this data base, the researcher can determine which particular product is most acceptable and which is least acceptable, as well as the range of acceptance and degree of difference.

F. Ratings of the Self-Designed Ideal Product

Magnitude estimation, just like any other form of intensity scaling, allows the panelist to rate the profile of the "ideal product", using the same scales that the panelist used to rate actual products. The panelist has an opportunity to "design a product", Table 16 shows the sensory attribute profile of the ideal product. Note that although there is not a systematic variation in the actual products, the profile of the ideal product can still be anchored in the range of products tested. On some attributes panelists may assign the "ideal product" a profile value higher or lower than any product achieved in the actual test (e.g., buttery). If the test includes a sufficiently wide range of products and yet the ideal product is beyond the limits tested, this discrepancy indicates that the attribute is probably another way of stating "liking". (For instance, the ideal level of "rich chocolate flavor" is generally higher

Table 15
QUESTIONNAIRE

PRODUCT EVALUATION

Name: _____ I.D.#: _____

Attribute	Product Ratings		
	Product #	Product #	Product #
	<u>467</u>	<u>641</u>	<u>322</u>
1. Degree of purchase intent	____	____	____
2. Overall acceptability <u>(Liking/Disliking)</u>	L/D | How much	L/D | How much	L/D | How much
3. Spreadability	____	____	____
4. Strength of buttery flavor	____	____	____
5. Degree of creamy taste	____	____	____
6. Degree of sweetness	____	____	____
7. Degree of saltiness	____	____	____
8. Smoothness of texture	____	____	____

than any chocolate product would achieve in a chocolate product test. Rich, chocolate flavor is really not a true sensory attribute, but combines sensory and acceptance characteristics.) The attribute "buttery" falls into the same category of unachievable ideal levels.

G. Relating Sensory Attributes to Liking

Scale values for attribute levels can be related to scale values for liking to reveal how sensory attribute levels co-vary with overall liking. One can relate the attribute level to liking to develop a curve such as that shown in Figure 6. That data plot (using a fitted curve) shows approximately how liking varies with sensory attribute level. If the curve is steep, then liking changes moderately with changes in the sensory attribute. If the curve is flat, then liking changes less dramatically and, indeed, may not change at all. Furthermore, the sensory attributes interact to generate liking. For heuristic reasons it is easier (and more instructive) to treat each attribute separately. We are primarily interested in the best "guess" as to how liking of an attribute varies with the level of the attribute.

H. Measuring Relative Importance of Attributes by Magnitude Estimation

Table 16 also shows the ratings for relative importance of different sensory attributes. The panelists were instructed to scale importance directly, by numerical ratings. The panelists were able to rate importance, but for some attributes (e.g., saltiness) the ratings are relatively low, suggesting "low importance" of that attribute. In actuality, when panelists are asked to scale importance directly, they often do the following:

1. They can easily scale importance of good quality, nutrition, etc. However, most of these "image attributes" generate similar (and very high) levels of importance.
2. They have a difficult time scaling the relative importance of sensory attributes. "How important is saltiness?" is a quesion that is hard to answer. Does the question refer

Table 16
DATA BASE — MAGNITUDE ESTIMATES OF MARGARINES ON ATTRIBUTES

	Buttery blind	Spreadable blind	Creamy taste blind	Smoothness blind	Sweetness blind	Saltiness blind	Like total blind	Like[a] User/ Q blind	Like[b] User/ R blind
A	79	112	89	106	74	76	63	65	61
B	79	100	82	99	70	73	39	59	19
C	65	81	70	84	63	61	37	37	37
D	60	105	78	93	64	52	30	29	31
E	63	95	75	93	65	66	28	9	47
F	53	102	74	95	61	65	12	20	4
G	50	100	65	90	55	63	9	10	8
H	46	103	62	92	50	47	−18	−23	−13
Ideal	104	104	76	102	70	58			
Importance	125	116	113	112	106	58			

[a] Users of Brand Q.
[b] Users of Brand R.

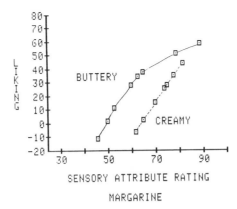

FIGURE 6. *Fitted* curve relating attribute level (e.g., buttery or creamy) to overall liking. The fitted curve shows the underlying relation between attribute level and liking. The relation can show curvature, suggesting an optimal sensory level. The steeper the curve, the more small changes in the sensory attribute lead to large changes in liking.

to "saltiness" per se or to "being right on target with the right level of saltiness of margarine"? The question is ambiguous, and panelists resort to scaling the ideal level of saltiness that they desire.

VI. SPECIFIC BENEFITS OF USING MAGNITUDE ESTIMATION

Many of the foregoing analyses can be easily performed with other scales such as fixed-point category scales. (The analysis is far more difficult, but still possible, using rank order scales.) Why, then, use magnitude estimation for either scientific research or applied sensory analysis when other scaling procedures are equally capable of generating good data?

A. Scale Quality

In scientific research, a ratio scale is the *summum bonum* of measurement. Even though a fixed-point category scale would probably work as well, ratio properties are preferred by scientists.

B. Applicability in Nonnumerical Measurement

Numbers (assigned by magnitude estimation) need not be the only units of measurement for ratio scales. Early in the research on scaling, Stevens and his colleagues recognized that panelists could adjust the perceived intensity of other continua besides numbers to match the perceived intensity of the stimulus. Continua that were adjusted included line length, force of handgrip, and noise.[28] For instance, the panelist could adjust the sound pressure of white noise so that the loudness matched the sweetness of sugar solutions.[29] The panelist could also adjust the size of numbers so that the size of the numbers matched the sweetness of the sugar solutions. Both matching functions were power equations, written below:

$$\text{Sound pressure} = k_1(\text{sugar concentration})^M$$

$$\text{Number (M.E.)} = k_2(\text{sugar concentration})^N$$

Table 17A
SUMMARY OF ANALYSIS OF VARIANCE: BEVERAGES[30]

	Full panel — N = 30				Half panel — N = 15			
Source	**DF**	**SS**	**MS**	**F**	**Source**	**DF**	**SS**	**MS**	**F**

				Magnitude Estimation[a]				
Foods	4	6.36	1.59	19.11	Foods	4	4.93	1.23	12.89
Panelist	29	4.67	0.16	1.94	Panelist	14	2.53	0.18	1.90
Error	116	9.70	0.08	—	Error	56	5.36	0.10	—
Total	149	20.68	—	—	Total	74	12.82	—	—

				Nine-Point Scale				
Foods	4	97.1	24.3	12.84	Foods	4	55.8	14.0	5.95
Panelist	29	113.6	3.9	2.07	Panelist	14	82.3	5.9	2.51
Error	116	219.3	1.9	—	Error	56	131.4	2.3	—
Total	149	430.0	—	—	Total	74	269.5	—	—

[a]   Not postmultiplied by 100.

Combining these two equations generates a prediction of the direct match of number to sound pressure (viz., magnitude estimation of loudness). By transitivity:

$$\text{Number (M.E.)} = k_3(\text{sound pressure})^{N/M}$$

The transitivity shown above, and the high predictability of matching functions is not possible if the scale is a category scale. It is only possible if the scale is a ratio scale.

As a consequence of this demonstration of transitivity, it appears that measurement by matching continua other than numbers to perceived sensory magnitude generates a ratio scale and not an interval or category scale.

C. Sensitivity to Differences

Magnitude estimation may yield somewhat better discrimination among stimulus than do category scales. Various studies comparing the two procedures have suggested that indeed the open-ended scaling does produce greater discrimination if this discrimination is measured by either the t-test for differences between means or the analysis of variance for differences among several products.

Table 17 (A and B) presents the results of analyses of variance, comparing normalized magnitude estimates and category scales for liking. Note that in most cases the magnitude estimation scale generates a higher F-ratio.[30]

D. Improved Ability to Generate Relations Between Variables

If a researcher is interested in modeling relations between physical measures and sensory or acceptance reactions, the magnitude estimation scale often generates better fits of the equation to the data, whereas limited category scales provide poorer fits.[30] This improvement in fitting the equation occurs because the category scale is limited and may force the average ratings towards the center of an already limited scale. This gravitation towards the center constrains the scale range (viz., the range of the dependent variable), making it more difficult to uncover a robust function. In contrast, the magnitude estimation scale does not force the ratings towards the center in such a dramatic fashion, because the magnitude estimation scale is broader. (The regression towards the mean occurs more clearly with category scales

Table 17B
SUMMARY OF ANALYSIS OF VARIANCE: CAKE[30]

Full panel — (A) N = 30				Half panel — (B) N = 15				
Source	DF	SS	MS	F	Source	DF	SS	MS	F

Magnitude Estimation[a]

Source	DF	SS	MS	F	Source	DF	SS	MS	F
Foods	4	6.10	1.52	6.35	Foods	4	5.90	1.48	4.92
Panelist	29	25.58	0.88	3.70	Panelist	14	14.58	1.04	3.47
Error	116	27.84	0.24	—	Error	56	16.80	0.30	—
Total	149	59.52	—	—	Total	74	37.28	—	—

Nine-Point Scale

Source	DF	SS	MS	F	Source	DF	SS	MS	F
Foods	4	172.84	43.20	3.36	Foods	4	188.4	47.1	2.22
Panelist	29	754.20	26.01	2.03	Panelist	14	380.8	27.2	1.28
Error	116	1482.3	12.78	—	Error	56	1187.7	21.21	—
Total	149	2409.3	—	—	Total	74	1756.9	—	—

[a] Not postmultiplied by 100.

because panelists are afraid of running out of numbers at either the high or the low end when they use scales with a limited range and few scale points.) Panelists act conservatively by maintaining their scale values in the midrange until confronted with a stimulus that is undeniably at the top or the bottom end of the scale. Only then do the panelists feel comfortable assigning the extreme stimuli ratings at the ends of the scale.

VII. LIMITS OF THE MAGNITUDE ESTIMATION METHOD

The foregoing section dealt with some of the benefits of using the scale. However, there are certain limitations to the scale, primarily in terms of field execution and analysis.

A. Orientation Time

Magnitude estimation requires orientation. As noted above, some researchers opt to orient the panelists rapidly, simply instructing them to match numbers to stimuli. Other researchers choose to have the panelists undergo a more rigorous training. More complex, more rigorous training takes more time. In field work, especially that involved in the evaluation of products, the benefits of magnitude estimation may be offset by the increased time required to learn how to scale. Evaluation of the trade-off between increased sensitivity and increased field work effort is up to the individual researcher.

B. Analysis Issues

Statisticians like data to behave perfectly. Category scale data do not behave perfectly, because the variability of the ratings is higher in the middle than at the extremes of the scale. For magnitude estimation, statisticians find that often the scale values are not perfectly normally nor log normally distributed. The magnitude estimates may not distribute log normally, even after normalization. The puristic statistician will reject magnitude estimation based upon this lack of normality. (However, the same statistician ought to reject the category scale as well because of nonnormality of the distribution at the upper and lower extremes of the scale).

VIII. AN OVERVIEW

Magnitude estimation as a scaling technique has become increasingly popular among scientists interested in the mechanisms of perception. The procedure does generate scales that appear to conform to the requirements of a ratio scale, despite the arguments to the contrary. As a procedure for obtaining a dose response curve relating sensory intensity to physical level the method has been proven reliable and useful. Indeed, one is hard pressed today (1988) to find many scientific papers dealing with perceptual magnitude that do not at least reference the method (even if to disagree with the approach). As such, in the scientific community, the method is highly accepted.

Among applied researchers, especially those dealing with consumer evaluation of packaged goods, magnitude estimation procedures have gained attention and sparked interest. The method clearly works in application and generates valuable information. Furthermore, because the scale generates validated ratio properties, it is useful as a measure of relative magnitude (for example, comparison of the perceived efficacy of two mouthwashes or the fragrance intensity of two fragrance submissions). The key problem with magnitude estimation appears to be the difficulties encountered in instructing panelists to use the scale. Those difficulties increase the time in the field (at the point of data collection) and thus increase the cost to run the study. For business-oriented research, the increased cost is certainly a factor to be weighed against the additional benefits of the scale. If increased sensitivity and better functional relations are of interest to the applied researcher, then certainly the magnitude estimation approach deserves consideration.

REFERENCES

1. **Stevens, S. S.,** On the theory of scales of measurement, *Science,* 103, 677, 1946.
2. **Guilford, J. P.,** *Psychometric Methods,* McGraw-Hill, New York, 1954.
3. **Stevens, S. S. and Poulton, E. C.,** On the halving and doubling of the loudness of white noise, *J. Acoust. Soc. Am.,* 329, 1955.
4. **Beebe-Center, J. G.,** Standards for use of the GUST scale, *J. Psychol.,* 28, 411, 1949.
5. **Woodworth, R. S.,** *Experimental Psychology,* New York, Holt, 1938.
6. **Stevens, S. S.,** Mathematics, measurement, and psychophysics, *Handbook of Experimental Psychology,* Stevens, S. S. Ed., John Wiley & Sons, New York, 1949, 1.
7. **Fechner, G. T.,** *Elemente der Psychophysik,* Breitkopf und Hartel, Leipzig, 1853.
8. **Stevens, S. S.,** On the brightness of lights and the loudness of sounds, *Science,* 118, 576, 1953.
9. **Stevens, S. S.,** On the psychophysical law, *Psychol. Rev.,* 64, 153, 1957.
10. **Marks, L. E.,** *Sensory Processes: The New Psychophysics,* Academic Press, New York, 1974.
11. **Stevens, S. S.,** *Psychophysics: An Introduction To Its Perceptual, Neural and Social Prospects,* John Wiley & Sons, New York, 1975.
12. **Engen, T. and Levy, N.,** The influence of standards in constant sum methods, *Percept. Mot. Skills,* 5, 193, 1955.
13. **Stevens, S. S.,** personal communication, 1968.
14. **Ekman, G.,** Two generalized ratio scaling methods, *J. Psychol.,* 45, 287, 1958.
15. **Stevens, S. S.,** *Psychophysics: An Introduction to its Perceptual, Neural, and Social Prospects,* John Wiley & Sons, New York, 1975, 30.
16. **Stevens, S. S.,** *Psychophysics: An Introduction to its Perceptual, Neural, and Social Prospects,* John Wiley & Sons, New York, 1975, 15.
17. **Moskowitz, H. R.,** The new psychophysics and cosmetic science, in *Cosmetic Science,* Vol. 2, Breuer, M., Ed., Academic Press, New York, 1980, 129.
18. **Stevens, S. S. and Guirao, M.,** The scaling of apparent viscosity, *Science,* 145, 1157, 1964.
19. **Moskowitz, H. R.,** Sweetness and intensity of artificial sweeteners, *Percept. Psychophys.,* 8, 40, 1970.
20. **Moskowitz, H. R.,** The sweetness and pleasantness of sugars, *Am. J. Psychol.,* 84, 387, 1971.

21. **Engen, T. and McBurney, D. H.,** Magnitude and category scales of the pleasantness of odors, *J. Exp. Psychol.,* 68, 435, 1964.

22. **Engel, F.,** Experimentelle Untersuchungen uber die Abhangigkeit der Lust und Unlust von der Reizstarke beim Geschmacksinn, *Pfluegers Arch. Gesamte Physiol.,* 64, 1, 1928.

23. **Moskowitz, H. R.,** Sensory intensity versus hedonic functions; classical psychophysical approaches, *J. Food Qual.,* 5, 109, 1981.

24. **Ekman, E. and Akesson, C. A.,** Saltiness, sweetness and preference. A study of quantitative relations in individual subjects, Report 177, Psychology Laboratories, Univ. of Stockholm, Stockholm, Sweden, 1964.

25. **Moskowitz, H. R.,** Sensory and hedonic functions for chemosensory stimuli, in *Chemical Senses and Nutrition,* Kare, M. R. and Maller, O., Eds., Academic Press, New York, 1977, 71.

26. **Moskowitz, H. R. and Klarman, L. E.,** The hedonic tones of artificial sweeteners and their mixtures, *Chem. Senses Flavor,* 1, 423, 1975.

27. **Moskowitz, H. R.,** Mind, body and pleasure: an analysis of factors which influence sensory hedonics, in *Preference Behavior and Chemoreception,* Kroeze, J., Ed., Information Retrieval, London, 1979, 131.

28. **Stevens, S. S. and Guirao, M.,** Subjective scaling of length and area and the matching of length to loudness and brightness, *J. Exp. Psychol.,* 66, 177, 1963.

29. **Moskowitz, H. R.,** Intensity scales for pure tastes and taste mixtures, *Percept. Psychophys.,* 9, 51, 1971.

30. **Moskowitz, H. R.,** Utilitarian benefits of magnitude estimation scaling for testing product acceptability, in *Selected Sensory Methods: Problems and Approaches to Measuring Hedonics,* Special Technical Publication 773, Kuznicki, J. T., Rutekiewic, A. F., and Johnson, R. A., Eds., American Society for Testing and Materials, Philadelphia, 1982, 11.

Chapter 11

STANDING PANELS USING MAGNITUDE ESTIMATION FOR RESEARCH AND PRODUCT DEVELOPMENT

Ronald S. Leight and Craig B. Warren

TABLE OF CONTENTS

I.   Introduction 226

II.   Standing Panels for Magnitude Estimation 226
        A.   The Choice of the Panel...227
        B.   The Choice of Magnitude Estimation................................227
                1.   The Advantages of Magnitude Estimation 227
                2.   The Disadvantages of Magnitude Estimation.................227
III.   Recruitment and Selection of Panelists......................................227
        A.   Recruiting 227
        B.   Screening Tests...228
                1.   Screening for General Olfactory Acuity and Ability
                        to Follow Instructions...................................228
                2.   Other Screening Tests....................................231

IV.   Panelist Training...231
        A.   Intensity Scale Training 231
        B.   Hedonics Scale Training...233
        C.   The 13-Sampler Odor Set 234
        D.   The Full Solution Set (23 Samples) 235

V.   Data Collection and Analysis 235
        A.   Mark Sense Cards..235
        B.   The Sensory Data System (SDS) 236
                1.   Overview 236
                2.   Entry of Sample Identification (ID) Information 236
                3.   Data Card Entry 238
                4.   Data Analysis..238
                5.   Report Generation 239
        C.   Normalization Methods...239
                1.   Internal Standard Normalization 240
                2.   No-Standard Normalization — The Method of
                        "Averages" 241
                3.   Fixed-Scale Normalization 242
                4.   External Modulus Normalization...........................242
        D.   Considerations and Experimental Limitations in the Use
                of Magnitude Estimation and the Various Normalization Methods 242

VI.   Judging Panelist Performance 243
        A.   Panelist Ranking 243
        B.   Panelist Scales...245
        C.   Panelist Normalization Factors 245
        D.   Summary 247

VII.   Panelist Motivation 247

VIII. Conclusions 248

Acknowledgments 248

References 249

I. INTRODUCTION

This article is a case study of the recruiting, training, panelist motivation, and data analysis practices of the International Flavors and Fragrances (IFF) Sensory Testing Center. It is primarily a "how-to" article that brings together some well-known, some not-so-well-known, and some original techniques for using and managing standing panels. Sensory panel screening and training methods are discussed, along with some practical warnings for the novice practitioner. Finally, this article deals primarily with panels for fragrance evaluation and magnitude estimation.

Section II discusses the pros and cons of a standing panel and magnitude estimation scaling. Sections III and IV deal with the recruiting and training of panelists. Section III also includes the distribution of subject scores for the screening test, and Section IV provides the results from one of the training exercises. Section V describes the data collection and analysis system. It includes a discussion of normalization methods and some of their limitations and a discussion of some potential problems in the use of magnitude estimation. Finally, the last two sections (VI and VII) describe some of the techniques used for panelist evaluation and motivation.

II. STANDING PANELS FOR MAGNITUDE ESTIMATION

The IFF Sensory Testing Center program originated from the customer's need for fragrances with demonstrable functional properties, such as substantivity and malodor masking. Fragrance *substantivity* is defined as a perceptible odor intensity on a surface such as skin, cloth, or hair after some defined processing and/or a specific period time. *Malodor masking* is defined as the ability of a fragrance to blend with and modify the perception of a particular malodor in a way that improves the overall hedonics of the mixture without increasing the overall perceived intensity.

There are a variety of useful psychophysical methods for determination of odor intensity and blending.[1-3] Some of these are ranking, paired comparison, category scaling, and various descriptive methods. These methods are used extensively by the consumer products industry, primarily at the food product companies. After evaluation of the classical methods, magnitude estimation was chosen because of the need to test thousands of ingredients on a routine basis and to predict the properties of formulations from the properties of these ingredients.

A. The Choice of the Panel

The use of sensory judgments for the creation and selection of products is certainly not new to IFF. Perfumers and flavorists make these judgments all the time. They have an extensive knowledge of fragrance and flavor ingredients and how to put them together. However, those same skills which make the expert excel in the creation of flavors and fragrances make it difficult for him to make simple global judgments of intensity and hedonics. The expert gets too involved in the fine details of the fragrance and has a tendency not to see the forest for the trees.

Another factor in the selection of the type of panel is cost. Neither the expert panel nor a trained ''lay'' descriptive panel is a cost-effective choice for the wide diversity of products to be evaluated and the narrow scope of judgments required. The ''one-use'', ''naive'' consumer panel is not cost-effective either, due to the large number of tests needed.

A standing experienced, ''lay'' panel provides a cost-effective, global measure of odor intensity. Properly selected and trained quantitative sensory panels become an analytical tool for the determination of odor intensity. Finally, the consumer evaluation of a product is more like that of the quantitative sensory panel than it is like that of the perfumer or expert descriptive panel.

B. The Choice of Magnitude Estimation

1. The Advantages of Magnitude Estimation

Magnitude estimation eliminates one of the principle drawbacks of category scaling: end effects.[4] End effects produce a number of biases. The one of most concern arises when the sample set contains a wide diversity of stimuli. If the panelists are not familiar with the gamut of experimental sensations, they may assign an early sample to a category which is either too large or too small for accurate evaluation of the subsequent samples. As a result, they run out of categories. Samples which more properly should be assigned to different categories are assigned to the same category. In principle, magnitude estimation provides an infinite number of categories in either direction from any value.

Magnitude estimation panels can be only two to three times larger than those used for expert or descriptive evaluations. Panels of 15 to 25 trained panelists produce data of adequate precision and reproducibility.

A final advantage is flexibility. Once trained in magnitude estimation, panelists use the method effectively on tasks ranging from the high intensity of dose-response and malodor-masking experiments to the extremely low intensity of cloth substantivity and 12-hr skin substantivity experiments. They adapt easily to stimulus presentations that vary from solutions in jars to soap bars to cloth to hair swatches to forearms to 7-gal garbage cans. In most cases, one or two sentences provide sufficient instruction for new types of experiments.

2. The Disadvantages of Magnitude Estimation

The disadvantages of using magnitude estimation lie primarily in the design and the analysis of the experiments. Magnitude estimation requires either a tedious normalization or the careful selection of an external modulus (Section V.C). The mathematics and statistics of normalization impose minimum sample set and panel sizes (Section V.D). Finally, efficient use of magnitude estimation requires extensive computerization (Sections V.A and V.B).

III. RECRUITMENT AND SELECTION OF PANELISTS

A. Recruiting

Approaches to finding panelists range from blind phone-calling to bulletin board notices and newspaper advertising to using a professional recruiter. Depending upon your screening criteria, you will need to recruit from four to ten people for every position on your panel.

A combination of approaches can generate a long list of candidates. The use of a telephone answering machine allows people to call at any hour. A respondent can leave his name, address, and phone number and receive detailed information by mail. One can anticipate that 30% of the initial respondents will call back after receiving the detailed program description. Once the initial panel group has been established, word of mouth produces a substantial waiting list of people interested in participating.

B. Screening Tests

The screening exercises depend upon the program objectives.[5] Among the potential screening criteria could be above-average olfactory acuity, above-average ability to discriminate among common odorants, motivation, and ability to follow directions. More specific criteria should be developed to meet the needs of the program.

1. Screening for General Olfactory Acuity and Ability to Follow Instructions

A useful test of olfactory acuity, discrimination, and ability to follow instructions consists of eight groups of three blotters.[6] Within each group, the three blotters may represent one, two, or three materials. The panelist is given a ballot containing eight triangles (Figure 1). The panelist is instructed (Figure 2) to put an '' = '' on the leg of the triangle that connects similar smelling blotters and and ''X'' on the leg of the triangle connecting blotters which have different odors. The panelist gets one point for each triangle leg incorrectly scored. Therefore, the best possible score is 0 and the worst is 24.

The odorants (Table 1) are chosen primarily for their familiarity to the average person. Another important consideration is to avoid materials which are rapidly fatiguing, such as musks and ionones. The relative difficulty of each comparison is seen in the percentage of incorrect responses (Table 2). The most difficult discriminations are those between Grisalva (CAS 88683-93-6) and Fixateur 404 (CAS 6790-58-5) and between lime and bergamot (Groups 2 and 8). The easiest discriminations are between spearmint and peppermint and among bay, spearmint, and lavandin (Groups 6 and 7).

Since 19 of the 24 responses should be ''different'', it is important to ask whether candidates are succeeding by simply answering ''different'' for all comparisons. The mean percentage for failing to detect true differences is 39.1%. This is the average percent wrong for all comparisons which are truly different (Table 2). The corresponding mean for incorrectly indicating a difference when none exists is 23.5%. This is a significant difference ($p < 0.02$, two-tailed t-test). Under this set of instructions, people are more likely to fail to detect a true difference than to report detecting a difference when there is none.

The difficulty of this test can be adjusted either through the choice of the materials in each triangle, or through the selection of the passing score. Using the 342 scores which have been collected over a 5-year period (Table 3), the pass/fail criteria for this test can be adjusted to accept any proportion of the population. In addition, 117 candidates have taken this test or an alternate version of it more than once. (The alternate version consists of the same eight groups of three blotters presented in a different order.) The average difference between scores for an individual is 2.6 (Standard deviation = 2.1). With this knowledge, the following strategy can be employed to accept the upper 50% of the population:

Score	Result
0—5	Pass
6—8	Retest — pass if retest score < 9
9—24	Fail

Retesting the middle group (scores of 6 to 8) compensates for the variability of the test

NAME: _____   **SET NUMBER:_____**

BALLOT FOR

OLFACTORY DISCRIMINATION

TRIANGLE TESTS

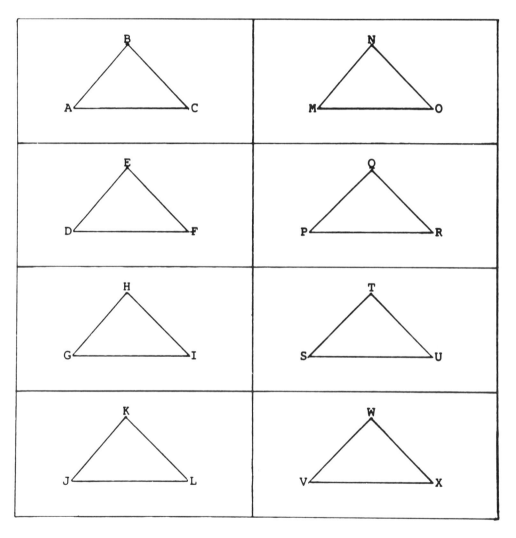

Date: _____   Number Wrong: _____

FIGURE 1.   Panelist screening ballot.

You have been given a number of sets of three blotters which have been dipped in perfume ingredients. Each set of blotters is different and may consist of:

1. Blotters of three different ingredients;
2. Two blotters with the same ingredient and one with a different ingredient; or
3. Three blotters of the same ingredient.

Work with one set of blotters at a time. You may sniff the blotters as many times, in any order and for as long as you feel may be necessary. This is not a speed test. However, first impressions are often the most accurate.

You have been given a ballot with a number of triangles on it. The number or letter at the corner of each triangle corresponds to a number or letter on a blotter. To record your evaluations, simply place an (X) between samples which are different and an (=) between samples which smell the same.

Example:

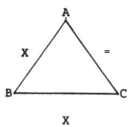

Here, Sample A smells the same as Sample C but both A & C are different from Sample B.

Note: If samples are different, they will be different odors not different intensities of the same odor.

FIGURE 2.   Instructions for screening triangles.

Table 1
IFF ODOR DISCRIMINATION TEST —
SET # 1

Group	Letter designation	Name
1	A	S. A. pettigrain oil
	B	Distilled Italian bergamot
	C	Distilled Mexican lime
2	D	Fixateur 404
	E	Grisalva
	F	Fixateur 404
3	G	California lemon oil
	H	Distilled Mexican lime
	I	California lemon oil
4	J	Spanish rosemary
	K	Terpineol
	L	French sauge sclaree
5	M	California orange oil
	N	California orange oil
	O	Grapefruit oil
6	P	Spearmint
	Q	Spearmint
	R	Peppermint
7	S	Bay oil
	T	Spearmint
	U	Terpeneless lavandin
8	V	Distilled Mexican lime
	W	Distilled Italian bergamot
	X	Distilled Mexican lime

result. Additionally, anyone who can achieve an above-average score on this test twice within a 1-hr period will be an excellent panelist. This is a strenuous test. Even those who do well on it find the number of samples and overall intensity levels to be fatiguing.

2. Other Screening Tests

It may be desirable to add other screening procedures to the selection process. For example, a test for specific anosmias or for the ability to correctly rank the intensities of samples in a dilution series might be appropriate. The necessity for additional screening will depend upon the nature of the evaluations expected from the panel.

IV. PANELIST TRAINING

Training sessions consist of four segments: intensity scale training, hedonics scale training, 13-sample odor set and 23-sample odor set. The full training session takes approximately 2.5 hr. One person can easily train 10 to 15 people. The only limitations are the need to observe each person's work and the desirability of working up the data from the first odor set before starting the second. Larger groups can be handled if a large enough staff is available to circulate and to answer questions on an individual basis.

A. Intensity Scale Training

Training begins with the magnitude estimation of the area of a series of geometric shapes. The candidates are instructed on the properties of an intensity scale and the principles of magnitude estimation. The presentation contains the following information:

Table 2
ERROR RATES IN THE 24 COMPARISONS IN
ORDER OF DECREASING DIFFICULTY

Group	Samples[a]	Percent wrong[b]
2	Fixateur 404 (D) vs. Grisalva (E)	59.9
	Fixateur 404 (F) vs. Grisalva (E)	61.4
	Fixateur 404 (D) vs. Fixateur 404 (F)	31.0
8	Lime (V) vs. bergamot (W)[c]	55.3
	Lime (X) vs. bergamot (W)	58.2
	Lime (V) vs. lime (X)	19.3
1	Pettigrain (A) vs. bergamot (B)	54.4
	Pettigrain (A) vs. lime (C)	37.1
	Lime (C) vs. bergamot (B)[c]	33.3
3	Lemon (I) vs. lime (H)	44.2
	Lemon (G) vs. lime (H)	38.3
	Lemon (G) vs. lemon (I)	23.4
5	Orange (M) vs. grapefruit (O)	41.8
	Orange (N) vs. grapefruit (O)	43.9
	Orange (N) vs. orange (M)	21.9
4	Rosemary (J) vs. terpineol (K)	26.6
	Rosemary (J) vs. sauge sclaree (L)	30.7
	Terpineol (K) vs. sauge sclaree (L)	31.0
7	Bay Oil (S) vs. spearmint (T)	24.8
	Bay Oil (S) vs. terpeneless lavandin (U)	22.2
	Spearmint (T) vs. terpeneless lavandin (U)	28.6
6	Spearmint (P) vs. peppermint (R)	25.2
	Spearmint (Q) vs. peppermint (R)	25.4
	Spearmint (P) vs. spearmint (Q)	21.9

[a] Letters in parentheses correspond to those in Table 1.
[b] Proportion of incorrect responses in 342 evaluations.
[c] Note context effect: Group 1 vs. Group 8.

1. An intensity scale starts at zero which means nothing is there; no intensity.
2. The scale is open-ended; it has no upper limit.
3. The method is based on a ratio principle. If a property is twice as large, it gets a number twice as large. If a property is half the size of a previous sample it gets a number half as large. Several examples are presented.
4. There are no right or wrong answers, only personal evaluations. The researcher is not interested in the panelist's neighbor's opinion or a group consensus, only their individual evaluation of each sample.

Following this, a randomly sorted collection of six rectangles, six circles, and six triangles of different sizes is presented. The figures range in size from 2.0 to 216.4 cm². Each figure is centered on an 8.5 × 11-in. page and identified with a five-digit random number. The 72.4-cm rectangle is presented as the first page of each booklet. A booklet and an answer ballot are distributed to each candidate. The session continues with the following instructions:

Our first exercise will be to estimate the areas of a number of shapes. Please write your name on the answer sheet. On the first page of the booklet in front of you, you will find a five-digit identification number; please write it on the first line of your ballot. Now choose some number between 30 and 50 which you feel best represents the area of this figure. Write this number down on your ballot.

Now turn the page. Write down the identification number of this figure. Now without looking back, estimate the area of this figure by writing down a number which best represents the ratio of this area to that of the previous

Table 3
DISTRIBUTION OF SCORES FOR THE IFF ODOR DISCRIMINATION
TEST — # 1

Score[a]	Frequency	Percent	Cumulative percent
0 —	0	0.00	0.00
1 — x	2	0.58	0.58
2 — xx.	5	1.46	2.05
3 — xxxxxxxxx.	19	5.56	7.60
4 — xxxxxxxxx.	19	5.56	13.16
5 — xxxxxxxxxxxx.	25	7.31	20.47
6 — xxxxxxxxxxxxxxxxxxxx.	41	11.99	32.46
7 — xxxxxxxxxxxxxxxx	34	9.94	42.40
8 — xxxxxxxxxxxxxxxxxx.	39	11.40	53.80
9 — xxxxxxxxxxxxxxxx	32	9.36	63.16
10 — xxxxxxxxxxxxxxxxx.	33	9.65	72.81
11 — xxxxxxxxxxxxxx	28	8.19	82.99
12 — xxxxxxxxxxxx.	23	6.73	87.72
13 — xxxxxxxxx.	17	4.97	92.69
14 — xx.	5	1.46	94.15
15 — xxxx.	9	2.63	96.78
16 — xxx	6	1.75	98.54
17 — .	1	0.29	98.83
18 — .	1	0.29	99.12
19 — .	1	0.29	99.42
20 —	0	0.00	99.42
21 —	0	0.00	99.42
22 — .	1	0.29	99.71
23 —	0	0.00	99.71
24 — .	1	0.29	100.00

[a] Score is the number of incorrect responses; "x" represents two scores; "." represents one score. If a candidate took this test more than once, all scores have been included.

figure. If this figure is twice as large as the previous figure, choose a number that is twice as large as that chosen for the previous figure. If this figure is a third of the area of the previous figure, choose a number that is a third of the one chosen for the previous figure.

Now turn the page. Write down the identification number of this figure. Now estimate the area of this figure by writing down a number which best represents the ratio of this area to that of the previous figure. If this figure is three times as large as the previous figure, choose a number that is three times as large as that chosen for the previous figure. If this figure is half the area of the previous figure, choose a number that is half of the one chosen for the previous figure.

Now turn the page. Continue doing one figure at a time until you have completed the booklet. Always compare the current figure to the immediately preceding figure. Do not turn back to look at the previous figure once you have turned the page. Just try to remember its size.

This exercise takes about 15 min to complete. During this time the instructor (and assistants) observe each candidate and give additional instruction where necessary. Typical results for this exercise have appeared elsewhere.[6,7]

B. Hedonics Scale Training

Training now proceeds to estimation of hedonics (like and dislike). Candidates estimate their like or dislike of the following words: flowers, sun, hate, worm, kiss, puppy, pollution, money, New York City, mud, perfume, murder, sex, cigar, spaghetti, rattlesnake, and love. This list of words was developed to cover a dynamic range of like and dislike.[6,7] Other lists of words could work just as well.

Candidates are told that in estimating hedonics they must make two decisions: (1) whether the stimulus is liked or disliked and (2) how much it is liked or disliked. The bipolar nature of the scale is illustrated. The principles of magnitude estimation are once again presented in the context of a bipolar scale. They are instructed that zero is the neutral point, representing neither like nor dislike.

The word list is distributed and the candidates are given the following instructions:

Using the hedonic scale we have just discussed, please rate how you feel about each of these words. If you like a word write an ''L'' next to it. If you dislike the word, write a ''D'' next to it. Then write down a number to indicate how much you either like or dislike the word. If you feel indifferent or neutral about the word write an ''N'' next to it and give it a zero for how much. Remember that the ratio of the numbers you use should reflect the ratio of the intensity of your feelings about each word.

This exercise also takes about 15 min to complete. Once again, the instructor (and assistants) observe each candidate, giving additional instruction where needed. Typical results for this exercise have appeared elsewhere.[6,7]

C. The 13-Sampler Odor Set

This exercise requires the estimation of both intensity and hedonics and is the candidate's first experience with doing both estimates on the same sample. Before they are given the samples, the candidates are trained in the use of the data entry cards (Section V, below). With this exercise, they put together all the pieces of a typical experiment.

The 13-sample odor set consists of five-step dilution series of Cinnamalva (CAS 4360-47-8) and of Isobutyl Quinoline (CAS 65442-31-1 [82%], CAS 67634-06-4 [15%]) and three 270-ppm *n*-butanol samples. The concentration series range in half-log steps from 2,000 to 200,000 ppm in diethyl phthalate. The butanol solutions are in distilled water. Samples are presented in 4-oz, wide-mouth jars. Each jar contains 5 g of solution.

Butanol was chosen as a reference standard for internal standard normalization of intensity (see below). It has a long history of use as an intensity standard.[8] Cinnamalva and Isobutyl Quinoline were chosen as representative of a range of intensities and both positive and negative hedonics.

The candidates receive a box of randomly ordered samples. They are instructed to select the samples in a random order, write down the identification code for each sample and then evaluate both intensity and hedonics. They are encouraged to make judgments based on first impression, but are told that they may sniff each sample as many times as they wish. However, once they have evaluated a sample, they may not go back to smell it again.

This exercise takes about 15 min. Once again the instructor (and assistants) monitor and correct each candidate's performance. Once the evaluations are completed, the candidates are given a 20- to 30-min break while the results of this exercise are evaluated (Table 4).

After the instructor has reviewed the results, each candidate is given a printout of his personal results and the summary of the group results. This serves a number of purposes. First, it illustrates and reinforces that their individual results are available to the staff for evaluation. Second, it provides a dramatic illustration of order effects and how they cancel out across the group. This reinforces the need to choose and evaluate samples in a random order. Third, it is an opportunity to build confidence. The candidates can know that even if their results are not ideal, they are doing well and that they will get even better with experience. Finally, it provides feedback on whether or not they are using the data cards properly.

At this point, most individuals are fully trained. However, going over the results of the solution set will expose the 1 person in 20 who is just not getting the idea. Generally, these are the individuals who for one reason or another cannot deal with numbers.

Table 4
TYPICAL RESULTS FROM 13-SAMPLE SOLUTION SET[a]

Solution	Concentration[b] (ppm)	Intensity[c]	Expected[d]	Hedonics [e]
Cinnamalva	2469	12.1 (1.26)	17.0	6.2 ± 8.4
	7407	21.6 (1.31)	24.6	14.4 ± 11.0
	22222	35.4 (1.33)	35.8	23.8 ± 11.1
	66667	48.6 (1.25)	52.1	24.3 ± 15.7
	200000	57.0 (1.23)	75.7	6.2 ± 22.4
Isobutyl quinoline	2469	25.0 (1.31)	19.3	− 33.9 ± 6.1
	7407	36.8 (1.24)	27.9	− 50.8 ± 8.0
	22222	43.8 (1.23)	40.5	− 66.8 ± 7.0
	66667	58.8 (1.22)	58.7	− 108.8 ± 14.6
	200000	58.3 (1.26)	85.1	− 101.4 ± 9.6

[a] Based on 18 candidates trained in two sessions.
[b] Concentration in diethyl phthalate.
[c] Geometric means are taken for all intensities. The numbers in parentheses are standard errors and should be interpreted as 1 + % error; e.g., 1.26 = 26% relative error.
[d] Calculated from the dose-response curve generated by a group of 15 experienced panelists. The curves are

    Cinnamalva:
        Intensity $= 1.19$ (concentration)$^{0.34}$
    Isobutyl Quinoline
        Intensity $= 1.37$ (concentration)$^{0.34}$

[e] Error ranges are standard errors.

D. The Full Solution Set (23 Samples)

After discussing the results of the 13-sample set, the next step is evaluation of a full solution set. This consists of 4 five-step dilution sets and three 270-ppm *n*-butanol samples. This is simply one more opportunity to practice the method and the use of the cards. For this exercise, the candidates are sent to the booths where they will normally be evaluating their samples. However, if this is not practical, this set can also be evaluated in the training area. Generally, there is no need to work up the data from this exercise. The samples used are generally left over from the active panels from that week.

At the end of this training session the panelists are ready to go to work. One can expect 10% of the people trained to drop out within the first couple of weeks and as many as 30% to drop out in the first 6 months. These are people who find that for one reason or another they are just not interested in continuing.

V. DATA COLLECTION AND ANALYSIS

A. Mark Sense Cards

One of the major problems encountered in large-scale sensory experiments is the time required to transcribe, proofread, and correct the raw data. In 1978, IFF developed a solution using ''mark sense'' cards and an optical mark reader. The custom designed card (Figure 3) has four fields: sample code, intensity estimate, hedonics estimate, and second intensity/ hedonics. The first two fields permit the entry of integers from 0 to 99999; the second two fields allow the entry of integers from − 99999 to + 99999. In addition to the fields provided for data entry, the cards are precoded with card, field, and end-of-card identifiers.

The Hewlett-Packard HP-7260A Optical Mark Reader (HP OMR) was selected as the best unit available for the purpose at the time. The HP OMR reads decks that have cards which are both punched and marked. This allows the preparation of special purpose cards

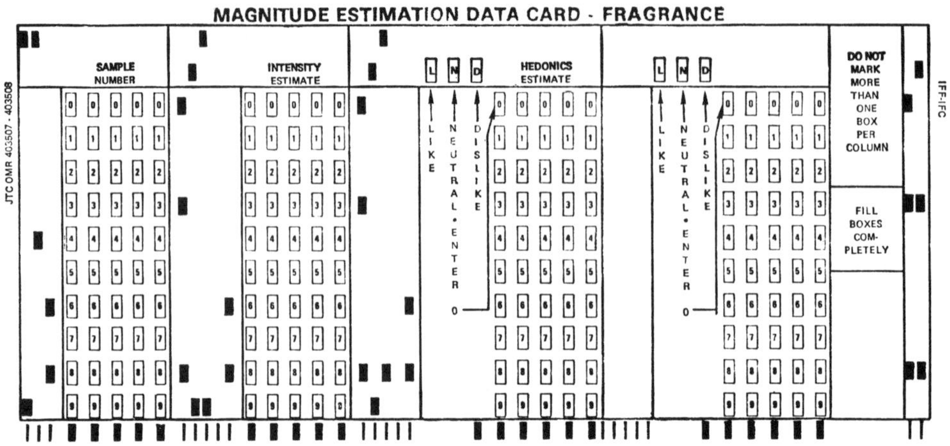

MAGNITUDE ESTIMATION DATA CARD - FRAGRANCE

FIGURE 3.   Magnitude estimation data card.

with a standard keypunch. One of these cards is the panelist identification or "WHO" card (Figure 4). This identifying card precedes each panelist's deck of data cards.

Two major types of errors must be controlled for in the use of mark sense cards: failure to detect intended marks and detection of unintended or incorrect marks. The first problem is almost completely eliminated by using No. 1 pencils and by training the panelists to make dark marks which go through the center of the box. The rate of failure to detect intended marks is less than 1/10,000. The second type of error can be handled through software/operator interaction. (See Section V.B.3 below.)

The major limitation of the mark sense card system is the cards themselves: the data must fit the card format. Careful design of the card and associated software minimizes the impact of this limitation. However, in designing a new system today, individual electronic data entry stations should be given serious consideration. These could be simple terminals connected to a central computer system, or a "lap" or "notebook" microcomputer for each panelist.

B. The Sensory Data System (SDS)

1. Overview

SDS is a modular, interactive data entry, analysis, and reporting system written in FORTRAN for the VAX® 11/780. It contains four main modules : (1) ID editor, (2) data card reader, (3) basic analysis and record keeping, and (4) report generator. Each of these sections can be operated independently. In addition, the files created by this program are structured such that any number of projects can be handled simultaneously. Once the project code is entered, the program leads the operator through a dialogue which determines which section to execute next.

2. Entry of Sample Identification (ID) Information

Before experimental data can be entered, the sample identification information must be provided. This is done through a crude, menu-driven ID editor. The ID editor provides seven commands: Add, Alter, Delete, Done, Help, List, and Print.

Add is executed automatically if SDS finds no data for the current project code. Otherwise it may be used at any time to add new samples to the file.

Alter allows any of the information for a single sample to be re-entered.

Delete deletes samples from the list.

Done exits from the ID editor.

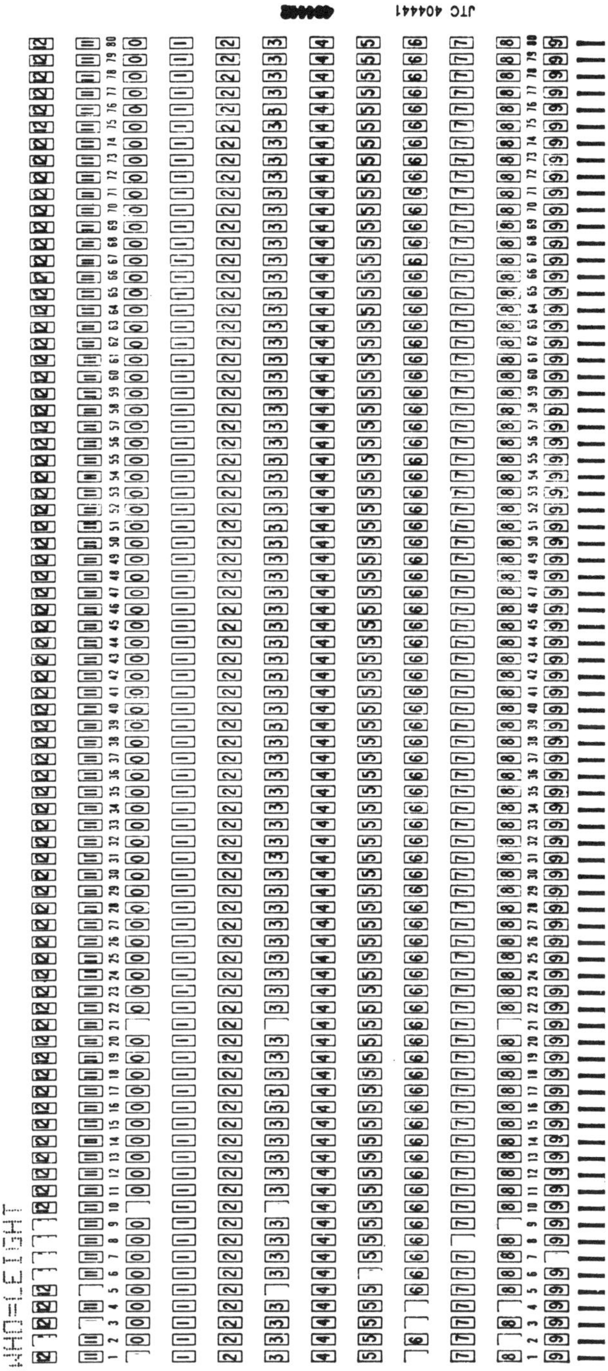

FIGURE 4. Panelist identification (WHO) card.

Help prints the list of valid commands and their definitions. This section also is executed any time an invalid command is entered.

List displays all of the current ID information on the terminal.

Print generates a hard copy of the ID list.

The ID file created by SDS can also be modified using the system's text editor.

3. Data Card Entry

Once the ID information is complete, SDS is ready to accept data. The first information required is the location of the data fields used. Any combination of one, two, or three fields of data can be collected on each sample. However, only fields 2 and 3 can contain bipolar data; the geometric mean is taken for field 1 and the arithmetic mean is taken for field 2.

Through each of the stages of data entry, all data read by the machine are validated through internal checks. This can be accomplished effectively only if the operator is running the program interactively. Whenever any error is detected, the program stops and requests that the operator make corrections. The program checks the validity of the card itself, each hand-coded character, and the sample ID number.

The most common type of card is the data card, identified by the code "ID = " in columns 1 through 3. On encountering this code, SDS checks for the card termination code. If this is not found in the proper location, operator intervention is required. This rare error generally occurs when a panelist inadvertently codes a machine control character in one of the data fields. It can also be caused by a badly printed card or by a machine error. Next, SDS checks each hand-coded field for nonnumeric data. Any stray mark, incomplete erasure, or inadvertent coding of two numbers in a single column will result in a nonnumeric or "nonsense" character. Any errors found at this point will be displayed on the screen for operator correction.

The next most common type of card is the "WHO = " or panelist name card. When this card is encountered, SDS begins checking the previous panelist's data for ID number errors. It compares the ID numbers read from the cards to those found in the ID file. Any unmatched or multiply matched ID numbers or cards are displayed at the terminal for operator correction. The most common errors are digit reversals and one-digit errors (e.g., 8 instead of 7 or 9). Once all the errors are corrected, the data are written to the data file.

The final legal card type is a completely blank card. This initiates the same error checking sequence as a "WHO = " card. In addition it signals the end of the deck.

While the potential exists for many errors, the actual error rate is low (Table 5). In a typical session, 86% of the panelists will make no errors and fewer than 5% of the panelists will make more than one error. On average, fewer than 1% of the cards read will have a detectable coding error.

4. Data Analysis

Once the data are entered, control is passed to the analysis and record-keeping section of the program. The operator may exit from the program or continue with the analysis. The program provides four different options for normalization of the data: (1) internal standards, (2) no standards, (3) no normalization, and (4) fixed factors. The first three are valid for all fields; the last is valid only for the third field (See Section V.C). Each field may be normalized by a different method, and the mean of field 3 may be taken either arithmetically or geometrically. As indicated previously, the default mean for field 1 is geometric and for field 2 is arithmetic.

Once the options have been chosen for each active field, the program calculates means and standard errors for each sample as well as the statistics on panelist performance. The panelist statistics include normalization factors, minimum, maximum, range, mean, and standard error for the raw scores in each field. In addition, their normalized intensity scores

Table 5
CARD ERRORS DETECTED DURING A
TYPICAL MONTH

Number of errors/set[a]	Number of data sets[b]	Percent of data sets	Cumulative percent of data sets
0	934	86.08	86.08
1	102	9.40	95.48
2	19	1.75	97.23
3	11	1.01	98.24
4	16	1.47	99.71
5	2	0.18	99.89
6	1	0.09	99.98
Total	1085		

Total number of cards read:	30,283
Total number of errors detected:	253
Error rate/card[c]:	0.84%

[a] A single card can have as many as four errors, one for each field (sample ID and up to three evaluations).
[b] A single data set consists of the cards from a single panelist for a single session.
[c] 100 × (Total number of errors/Total number of cards read).

are used to rank each panelist's performance (see below, Section VI). These results are written to an intermediate file which becomes the input for the report writer.

5. Report Generation

A menu-driven report generator allows the operator to request as many or as few reports as are necessary for each type of panel. Most, but not all, of these reports produce hard-copy output. Most also create data base files to be included in the master archive system. Many of these reports are highly specialized and provide additional detailed comparisons between predetermined samples or between the values in pairs of fields for a single sample. However, two reports are fundamental to each analysis: the raw data report and the compiled sample summary report (Figures 5 and 6).

C. Normalization Methods

Analysis of magnitude estimation data is typically a two-stage process. The first stage brings all of the panelists onto the same scale through some sort of ''normalization'' procedure. The panelists are not told what range of numbers to use; rather, they are told to choose numbers so that the ratio of these numbers reflects the ratios of their perceptions. This matching of ratios is the central element of the magnitude estimation method and must not be destroyed by the normalization process. Thus, normalization procedures are restricted to multiplication or division of the raw data by a correction factor. Addition or subtraction of correction factors would destroy the ratio scale attribute and thus are disallowed.

The second stage of the data analysis is determining the mean and standard error for each sample. It has been recommended that geometric means be employed with magnitude estimation data. This recommendation has generally been followed with unipolar scales. However, it is not possible to determine the geometric mean for a bipolar scale. This has not proved to be a problem.[9]

```
CONFIDENTIAL PROPERTY OF INTERNATIONAL FLAVORS & FRAGRANCES,INC.
SDS REVISION 0.99 (AUG-1985)                    9-MAR-86    11:24:31

DATA FOR: PANELIST 35            RANK:  1    CARD ERRORS:    0
PANEL: PN860307X                 INTENSITY RANK SUM:  164.0

NUMBER OF SAMPLES NORMALIZED= 15

INTENSITY 1  WAS NORMALIZED USING NO STANDARDS
             ---CONVERSION FACTOR USED WAS:    0.955
             ---RANGE:    65. -    150.
             ---AVERAGE=   102.3   S.E.=    5.97

HEDONICS 1   WAS NORMALIZED USING NO STANDARDS
             ---CONVERSION FACTOR USED WAS:    1.258
             ---RANGE:  -150. -      5.
             ---AVERAGE=   -72.5   S.E.=   11.82

INTENSITY 2  WAS NORMALIZED USING FIXED VALUES
             ---CONVERSION FACTOR USED WAS:    0.955
             ---RANGE:     0. -     75.
             ---AVERAGE=    33.7   S.E.=    6.80
```

COMPOUND OR FORMULA [PPM]	SAMPLE NUMBER	RAW INTEN 1	HEDON 1	INTEN 2	NORMALIZED INTEN 1	HEDON 1	INTEN 2
123456	(POS.CONTROL)						
2.00	65324	100.0	5.0	0.0	95.5	6.3	0.0
234567	ODORANT #1						
2.00	25814	125.0	-80.0	25.0	119.4	-100.6	23.9
2.00	28277	100.0	-80.0	20.0	95.5	-100.6	19.1
345678	ODORANT #2						
2.00	52464	125.0	-75.0	55.0	119.4	-94.4	52.5
2.00	62764	100.0	-100.0	50.0	95.5	-125.8	47.7
456789	ODORANT #3						
2.00	22708	85.0	2.0	0.0	81.2	2.5	0.0
2.00	83435	65.0	-10.0	10.0	62.1	-12.6	9.5
567890	ODORANT #4						
2.00	20166	100.0	-125.0	25.0	95.5	-157.3	23.9
2.00	86269	125.0	-100.0	70.0	119.4	-125.8	66.8
678901	ODORANT #5						
2.00	58447	75.0	-25.0	25.0	71.6	-31.5	23.9
2.00	87575	120.0	-85.0	70.0	114.6	-106.9	66.8
789012	ODORANT #6						
2.00	25966	90.0	-100.0	0.0	85.9	-125.8	0.0
2.00	43116	150.0	-150.0	30.0	143.2	-188.7	28.6
9999999	MALODOR(NEG.CONTROL)						
2.00	36345	75.0	-90.0	75.0	71.6	-113.2	71.6
2.00	50169	100.0	-75.0	50.0	95.5	-94.4	47.7

FIGURE 5.   Sample SDS raw data report.

1. Internal Standard Normalization[6,9]

Internal standard normalization works well for measurement of the intensities of flavors and fragrances. One of the advantages of using an internal standard is that it serves as a reference for linking the results from many independently run panels. The difficulty arises in picking an appropriate standard for a particular type of evaluation. Ideally, it should have an intensity that is close to the mean intensity of the complete sample set. At the very least, it should not have an intensity that is significantly greater than that of the most intense test sample, nor should it have an intensity that is significantly less than that of the weakest test sample. Violation of this requirement results in contrast distortion of the normalized intensity values.[9] A second requirement is that the standard should be presentable in a form which does not make it stand out from the test set, e.g., the standard should not be a liquid if the rest of the samples are solids.

```
CONFIDENTIAL PROPERTY OF INTERNATIONAL FLAVORS & FRAGRANCES,INC.
SDS REVISION 0.99 (AUG-1985)                    9-MAR-86    11:24:30

COMPILED DATA FOR PANEL: PN860307X
INTENSITY 1   WAS NORMALIZED USING NO STANDARDS
HEDONICS 1    WAS NORMALIZED USING NO STANDARDS
INTENSITY 2   WAS NORMALIZED USING FIXED VALUES
NUMBER OF PANELISTS:   22

COMPOUND
OR
FORMULA        SAMPLE   MEAN     STD ERR  MEAN     STD ERR  MEAN     STD ERR
       [PPM]   NUMBER   INTEN 1  INTEN 1  HEDON 1  HEDON 1  INTEN 2  INTEN 2
-----------------------------------------------------------------------------
123456       (POS.CONTROL)
       2.00    65324    104.4    1.07     -10.4    18.74    1.3      1.23
-----------------------------------------------------------------------------
234567       ODORANT #1
       2.00    25814    85.0     1.05     -64.9    10.82    16.7     1.41
       2.00    28277    79.4     1.06     -62.6    10.19    9.5      1.38
-----------------------------------------------------------------------------
345678       ODORANT #2
       2.00    52464    108.6    1.05     -110.0   7.23     3.4      1.44
       2.00    62764    110.2    1.04     -107.3   8.10     3.4      1.44
-----------------------------------------------------------------------------
456789       ODORANT #3
       2.00    22708    98.8     1.04     -63.0    13.54    3.2      1.40
       2.00    83435    96.8     1.05     -65.2    15.15    4.0      1.42
-----------------------------------------------------------------------------
567890       ODORANT #4
       2.00    20166    101.6    1.05     -117.9   5.83     3.4      1.43
       2.00    86269    106.1    1.04     -112.1   6.89     5.7      1.52
-----------------------------------------------------------------------------
678901       ODORANT #5
       2.00    58447    81.8     1.05     -60.1    11.74    7.4      1.41
       2.00    87575    84.8     1.04     -53.4    13.78    5.8      1.43
-----------------------------------------------------------------------------
789012       ODORANT #6
       2.00    25966    106.6    1.03     -95.9    10.58    3.1      1.41
       2.00    43116    103.8    1.04     -89.7    14.02    2.8      1.41
-----------------------------------------------------------------------------
9999999      MALODOR(NEG.CONTROL)
       2.00    36345    89.0     1.07     -98.3    6.16     80.2     1.09
       2.00    50169    76.6     1.06     -78.1    9.75     70.1     1.06
=============================================================================
THE GEOMETRIC MEAN IS CALCULATED
FOR ALL INTENSITY DATA.
```

FIGURE 6.   Sample SDS compiled sample summary report.

n-Butanol at 270 ppm works well as an internal standard for fragrance dose-response work.[8] Three butanol-in-water standards are included in a typical sample set comprised of 20 test samples. The sequence in which each panelist smells the samples and standards is random.

The correction or normalization factor for a particular panelist is the constant that will adjust the average of the perceived intensities for the butanol samples to 30. (The value of 30 was determined from a dose response curve for butanol and represents the mean unnormalized value given to a 270-ppm butanol/water solution.) In practice this factor is calculated by averaging a panelist's three values for butanol and dividing this average into 30. The panelist's test sample intensities are then multiplied by this ratio. Following the normalization process, the geometric mean and standard error for each sample are taken across all panelists.

2. No-Standard Normalization — The Method of "Averages"[6,9]

The method of averages can be used for any set of magnitude estimation data, provided the number of samples and panelists is sufficiently large (Section V.D, below). The key drawback to this method is that it does not allow direct comparison of magnitudes between panels. If such a comparison is needed, a reference sample must be included in all sessions which are to be compared.

The rationale for this method is fairly simple. All of the panelists have been exposed to the same total magnitude of stimuli, and should give the same total magnitude of response.

Therefore, one calculates a factor for each panelist which, when multiplied by each of his responses, produces a scale whose total magnitude is the same as that of every other panelist.

To accomplish this, one first totals the absolute values of each panelist's evaluations. Then the average total for the panel is calculated. Finally, the normalization factor for each panelist is the ratio of the average total to the individual panelist's total. Each panelist's data are multiplied by his normalization factor and the appropriate mean and standard error for each sample is calculated.

3. Fixed-Scale Normalization

When panelists are required to provide two different intensity ratings for each sample, another normalization method becomes possible. For example, in an odor-blending experiment, the first intensity might be the total sample intensity and the second the intensity of a particular odor character of the sample. Here, the number of 0 ratings for the second intensity can be quite large. When this occurs, normalization by the method of averages produces results which are both distorted and unreliable. However, if the panelists are instructed to rate the second intensity relative to the first, the second intensity can be normalized using the normalization factor calculated for the first.

4. External Modulus Normalization

It is sometimes impossible to design an experiment which allows postexperiment normalization of the data. An example of this is toothpaste evaluation. Here it is impossible for a panelist to evaluate enough samples in a single session to permit a valid normalization by either the internal standard or the method of averages (see Section V.D, below). In this case, an external modulus normalization is necessary. Each panelist is given a reference sample and told to assign its odor intensity a specific value, e.g., 100. Then the panelists rate the flavor intensity of the toothpaste by giving it a number which reflects its ratio to that of the modulus' odor intensity. The data collected in this manner require no additional normalization.

D. Considerations and Experimental Limitations in the Use of Magnitude Estimation and the Various Normalization Methods

There are a number of restrictions in the use of free-scaling magnitude estimation for flavor and fragrance evaluation which have not been explicitly recognized in the literature. These involve three intimately related factors:

1.   Sample set size
2.   Number of evaluations needed per sample
3.   Between and within set order effects.

Using the data collection and analysis system described here, it is possible to collect as many as 10,000 evaluations a week on a routine basis. When processing this many samples with a limited panel pool, one must be concerned about fatigue and loss of sensitivity. The maximum number of samples that a panelist can do in a single sitting (15 to 30 min) depends upon the nature of the task. For simple intensity and/or hedonics evaluations on moderately intense samples 20 to 30 samples can be done easily. However, an experiment which requires two different types of intensity evaluations on moderately strong, unpleasant samples (e.g., malodor masking experiments), the maximum number drops to about 15. In the extreme case, e.g., toothpaste evaluations, only one or two samples can be done at one time.

The number of samples which can be done in one sitting has a significant impact on both the experimental design and the choice of normalization method. For 20 or more samples, all normalization methods are applicable. However, the method of averages becomes less reliable as the number of samples decreases. Caution is recommended in using this method with fewer than ten samples. For smaller sample sets, the external modulus method is a better choice.

The number of panelists required is also an empirical question. IFF generally works with 20 to 24. For most exercises, 12 to 15 panelists should be considered a minimum. Panels using smaller numbers of panelists need to be highly trained to a common scale. This eliminates the need for normalization and reduces variability to a point where a small number of panelists can produce highly reliable results.[10]

The number of samples and the number of panelists is also affected by another issue.[11] Whenever a normalization factor is calculated, statistical theory requires the loss of a degree of freedom. When you have 10 samples and 15 panelists (149 degrees of freedom) the loss of 15 degrees of freedom from the analysis (one for each panelist) is not critical. However, with only 3 samples and 3 panelists (8 degrees of freedom), the loss of 3 degrees of freedom for normalization is a real problem. This reinforces the need to avoid normalization when degrees of freedom are restricted.

The final issue is order effects. Order effects within a set are well documented.[3] This can be dealt with primarily by instructing panelists to select their samples in a random order or by explicitly balancing the order of presentation. With 10 or more samples and 15 or more panelists, there is very little chance of any two panelists picking their samples in the same order. However, with fewer samples, some care must be taken to deliberately scramble the sample order between panelists or to explicitly counterbalance the sample sets and to control the order of evaluation.

Order effects are also a problem between sets. Just because the panelists can do one set of samples, this does not imply that they can do two sets of the same type in a row. One also cannot assume that because they can do set A and set B, the results are independent of the order in which the sets are presented. These are issues which must be addressed empirically. Panelists at IFF do as many as four sets of samples in a 2-hr period. However, careful consideration and experimentation have determined the types of experiments which are combined and the order in which they are done. Not surprisingly, it is better to do four different tasks or types of samples and to do the most difficult evaluations first.

VI. JUDGING PANELIST PERFORMANCE

A. Panelist Ranking

Panelists are told that there are no right or wrong answers, only their opinions. Despite this instruction there must be an objective reality, even if it is not directly measurable. Some measure of panelist performance against this reality would be of great value. For odor intensity and hedonics, there is no known way or measuring "reality" without using people. However, a suitably large group of people would provide an indirect measure of this "reality". Panelist ranking is a method of comparing the long-term performance of each panelist to that of the group.

This method was inspired by the nonparametric statistical technique known as Friedman Analysis of Variance by Ranks.[12] The rationale is as follows. Rank the normalized intensity ratings for a sample from highest to lowest. If all of the panelists are using the same criteria for rating intensity, their rank for any particular sample should be randomly drawn from a rectangular distribution of the values from 1 to the number of panelists. If, over a long period of time, a panelist's rank sum is consistently higher or lower than that of the rest of the group, he is doing something differently.

In practice, rank sums are converted into a ranking by first calculating the difference between an individual's rank sum for a panel and its expected value (Equation 1).

Table 6
PANELIST PERFORMANCE RANKING

	Normalized intensity score (rank)[a]				
	Panelist				
Sample	**1**	**2**	**3**	**4**	**5**
1	13 (2)	11 (3)	10 (4)	7 (5)	15 (1)
2	13 (2.5)	13 (2.5)	12 (4)	14 (1)	10 (5)
3	10 (1)	6 (4)	8 (3)	4 (5)	9 (2)
4	8 (3)	7 (4)	9 (2)	6 (5)	12 (1)
5	11 (3)	13 (1)	12 (2)	7 (5)	10 (4)
6	13 (2)	14 (1)	11 (3)	7 (5)	8 (4)
Rank sum	13.5	15.5	18	26	17
|Diff|[b]	4.5	2.5	0	8	1
Panelist rank	4	3	1	5	2

[a] Each entry is the normalized intensity value followed in parentheses by the rank of that value for that sample. 1 represents the highest intensity rating for that sample, 2 the second highest, etc.

[b] The absolute value of the difference between the panelist's rank sum and the expected value of 18. The expected value can be calculated either from equation (1) or as the average of the panelists' rank sums.

$$\text{Expected Rank Sum} = (1 + \text{number of panelists})/2 \times \text{number of samples} \qquad (1)$$

The absolute values of these differences are ranked from smallest to largest. The panelist with the smallest difference from the expected value is ranked number 1, the next smallest number 2, etc. In this manner, those panelists whose results are most representative of the group receive the smallest ranks and those whose data is least representative of the group the highest.

Table 6 illustrates the ranking process. The data are from a hypothetical panel with five panelists and six samples. The normalized values for each sample (rows) are first ranked from 1 (highest) to 5 (lowest). When two or more intensitites are the same (sample 2), they each receive an average rank. For example, the values 15, 13, 13, 13, 12 would be ranked as 1, 3, 3, 3, 5, respectively. Next the ranks for each panelist (columns) are summed. The absolute value of the difference of each rank sum from the expected value of 18 is ranked from 1 (smallest) to 5 (largest). In this example, panelist 3 is most representative and panelist 4 is least representative of the panel as a whole.

Although panelist 4 is the furthest from the "ideal", a closer look at the data suggests that this results largely from the evaluation for sample 2. This could be a misperception by the panelist or simply an order effect. It must be emphasized that a panelist's rank from any one panel session should not be considered a meaningful measure of long-term performance. Only the average rank over many sessions should be used as a tool for spotting aberrant panelists.

Periodic histograms of the average ranks of a group of panelists serve as one good way of spotting the aberrant members (Figure 7). Most of the 22 panelists have average ranks between 8 and 12. One panelist, with a rank or 14, stands out as being atypical. From speaking to this individual it was discovered that she had interpreted the instructions (Section IV) to mean that if she smelled nothing to give the sample an intensity of 0, otherwise to give it a number between 30 and 50! After retraining, this individual's performance became

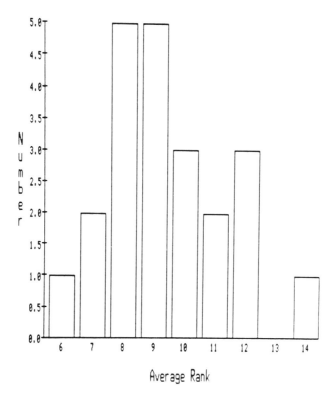

FIGURE 7. Histogram of average rank of 22 panelists. The y-axis is the number of panelists with that average rank based on approximately 120 sessions.

indistinguishable from that of the rest of the group. This is a typical outcome. Speaking with the aberrant panelists may not always reveal what they are doing differently, but it gets them thinking about what they are doing and how they are doing it. This generally resolves the problem.

Another good way of spotting aberrant panelists is to look at the histogram of each individual's ranks. There are three typical patterns (Figures 8 to 10). The first (Figure 8) is the "ideal" pattern, a uniform distribution of scores. The second (Figure 9) is the "atypical" performer. The scores are skewed toward higher ranks. This individual can generally benefit from counseling and retraining. The final pattern (Figure 10) is the "too-good-to-be-true" panelist. The results are skewed toward the lower ranks. This person may in fact be more representative of the group than the other panelists. However, he might be getting low ranks by not discriminating at all! Distinguishing between these two possibilities requires other techniques.

B. Panelist Scales

The major technique for identifying panelists who are not discriminating is to watch their scales. It is important to track raw scale minima, maxima, means, and standard deviations. Panelists who are using a restricted range of numbers are readily identified with this information. They can then either be retrained or dropped from the panel.

C. Panelist Normalization Factors

Ideally, all normalization factors should be 1, i.e., all panelists should gravitate to the same scale. In practice this is not true. Acceptably performing panelists can have average

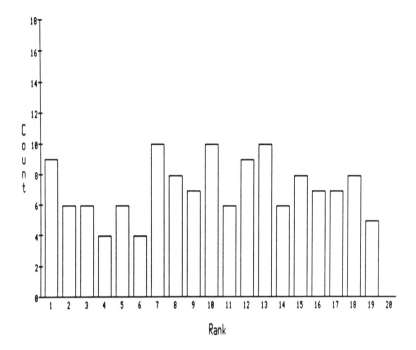

FIGURE 8.   Histogram of ranks for an ideal panelist. The y-axis is the number of times the panelist had that rank.

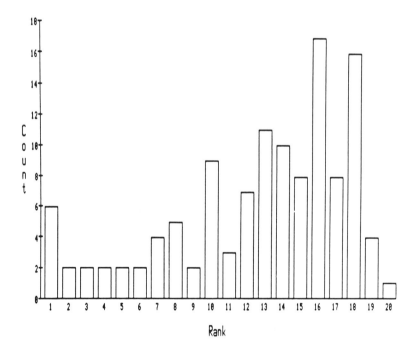

FIGURE 9.   Histogram of ranks for an anomalous panelist. The y-axis is the number of times the panelist had that rank.

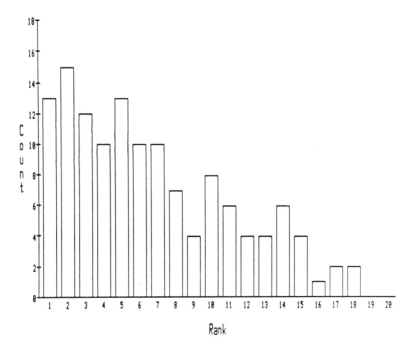

FIGURE 10. Histogram of ranks for a "too-good-to-be-true" panelist. The y-axis is the number of times the panelist had that rank.

normalization factors ranging from 0.2 to 3.0. The former represents someone who uses a particularly large scale, which generally is not a problem. The latter is someone who uses a particularly small scale. Whether this represents a problem can be evaluated using the two previously discussed techniques. However, even a panelist who has an "ideal" average normalization factor of 1.09 has factors for individual panels which range from 0.26 to 15.27.

Thus, like rank, on a panel-by-panel basis, panelist normalization factors are not necessarily an indicator of overall panelist performance. However, when plotted over the course of time, trends can be indicative. Of particular interest is any trend toward larger normalization factors. This would reflect a scale compression relative to the group and might represent a panelist who is becoming less discriminating.

D. Summary

Panelist ranking, scale statistics, and normalization factors all provide measures of panelist performance. However, neither the three together nor any one of them individually provides a clear discriminator between good and poor panelists. These techniques are primarily useful for pinpointing those panelists in need of attention. The panel manager must still use personal contact, good judgment, and good experimental design to determine whether a panelist is contributing valid data.

VII. PANELIST MOTIVATION

Panelist motivation is critical to maintaining a high degree of reliability in a standing panel. The motivational tools available fall into two major classes: material and psychological. The material rewards available are dependent on the company's position in the marketing chain and on whether employee or outside panelists are used. Material rewards range from direct monetary compensation to gift certificates and product gifts. The range of psychological techniques available are similar to those typically used for employee motivation.

For employee panelists who are participating as part of their job, direct monetary compensation for participation is probably not an option. On the other hand, with outside panelists this is probably the most effective means of maintaining interest and participation. How much panelists should be paid depends upon many factors: how many hours a week they will participate, how many trips they will make, how far they have to drive, how unpleasant the work is, etc. Whatever amount is decided on, after some length of time it will be perceived as inadequate. Therefore, other techniques are essential.

Product distributions depend upon what is available. At IFF, it is usually limited to anonymous fragrances. This reflects the company's policy of not identifying its customers and of not favoring one customer or potential customer over another. On the other hand, programs run by food and personal products companies should have no problem distributing samples of their products as incentives.

Psychological motivation is provided primarily by feedback. In most cases direct feedback about experiments is difficult. It is important to keep panelists as naive as possible about the nature and/or identity of the samples they are evaluating and the experimental design. Also, as they are continually reminded, there are no right or wrong answers. However, they can be praised for not making errors in filling out their cards. Each person's data are examined on a continuing basis, and panelists who are using restricted scales are counseled and retrained. Regular messages are posted on topics ranging from reminders about potential problems to notices of major sales based on their work.

The most important of all motivational techniques is to listen to panelist comments and to make sure they know that they are important. Aside from providing the data, these are the people who provide feedback, both through comments and through panel results themselves, on when the experiments are too difficult. Quarterly meetings are held for the whole group of panelists. This provides an opportunity to address specific problems and to provide some refresher training for everyone. It is an opportunity to discuss the state and future of the program. Even more importantly, it is a formal opportunity for them to air their complaints and discuss problems which they see in the program. It is also an opportunity just to get together and talk. This type of meeting is a tremendous morale booster.

Regardless of what material compensation is provided, it is essential to remember that panelists are people. The people must want to do the job and must want to do it well in order for the program to succeed.

VIII. CONCLUSIONS

While standing panels and/or magnitude estimation may not be the right choice for every need, they can be used productively for product development. At IFF, the system and methods discussed here meet the need.

ACKNOWLEDGMENTS

The authors would like to thank W. E. Brugger for converting and expanding R. S. L.'s orignial data analysis system so that it could be used on the VAX 11/780; L. T. Schreck for her contributions to the SDS, for training most of the panelists who have been through the Center, and for keeping the place running; and all those others whose technical and clerical assistance contributed to the development of the Sensory Testing Center.

REFERENCES

1. **Moskowitz, H. R.,** *Cosmetic Product Testing: A Modern Psychophysical Approach,* Marcel Dekker, New York, 1984.
2. **Kuznicki, J. T., Rutkiewic, A. F., and Johnson, R. A., Eds.,** *Selected Sensory Methods: Problems and Approaches to Measuring Hedonics,* Special Technical Publication 773, American Society for Testing and Materials, Philadelphia, 1982.
3. American Society for Testing and Materials, *Manual on Sensory Testing Methods,* Special Technical Publication 434, ASTM, Philadelphia, 1968.
4. **Moskowitz, H. R.,** Psychophysical and psychometric approaches to sensory evaluation, *Crit. Rev. Food Sci. and Nutr.,* 9, 41, 1977.
5. **American Society for Testing and Materials,** *Guidelines for the Selection and Training of Sensory Panel Members,* Special Technical Publication 758, ASTM, Philadelphia, 1981.
6. **Warren, C. B.,** Development of fragrances with functional properties by quantitative measurement of sensory and physical properties in odor, in *Odor Quality and Chemical Structure, American Chemical Society Symposium 148,* Moskowitz, H. R. and Warren, C. B., Eds., American Chemical Society, Washington, D.C., 1978, 57.
7. **Moskowitz, H. R.,** Utilitarian benefits of magnitude estimation scaling for testing product acceptability, in *Selected Sensory Methods: Problems and Approaches to Measuring Hedonics,* Special Technical Publication 773, American Society for Testing and Materials, Philadelphia, 1982, 11.
8. American Society for Testing and Materials, E-544: Standard practice for referencing suprathreshold odor intensities, in *1985 Annual Book of ASTM Standards, 15.07,* ASTM, Philadelphia, 1985, 31.
9. **Moskowitz, H. R.,** Magnitude estimation: notes on how, when, why and where to use it, *J. Food Qual.,* 1, 195, 1977.
10. **Levine, M.,** The use of magnitude estimation with parametric statistical methods in the subjective measurement of axillary odor presented to ASTM Committee E-18, Fort Lauderdale, Fla., December 1984.
11. **Pearce, J. H., Korth, B., and Warren, C. B.,** Evaluation of three scaling methods for hedonics, *J. Sensory Stud.,* 1, 27, 1986.
12. **Siegel, S.,** *Nonparametric Statistics for the Behavioral Sciences,* McGraw-Hill, New York, 1956, 166.

Index

INDEX

A

Ability to follow instruction screening test, 228—231
Absolute number matching, 196—197
Additives, sensory effects of, 23
Adhesiveness scale, 95
Aftertaste, 29, 33
American Society of Brewing Chemists (ASBC), 74, 82—83
Amplitude, 24, 29, 32
Analysis of variance (ANOVA), 36, 59, 162—163
  beverages, 220
  cake, 221
Appearance order, see Order of presentation
Aroma, see also Fragrance substantivity; Odor
  definition of, 27
  imagery of, 11
Aroma note index, 31
Aromatics
  complexity of, 33
  references for, 28
Arrangement test, 38
Assessor, see Panelists; Respondents
ASTM Ascending Method of Limits, 83
Attributes
  correlation coefficients for, 60
  definition of, 54—55
  measurement of relative importance of, 217—219
  physical magnitude and, 205
  relation of to liking, 217
  scale for, 34—36
  scoring of by good and poorly performing panels, 64—65
  selection of, 33—34
Averaging method, 241—242

B

Basic taste test, 38
Beer flavor terminology, 73—86
  arguments for, 74
  flavor reference substitutes, 84—85
  history of development of, 74
  principles of system of, 74, 80—81
  recommended descriptors for, 75—80
  reference standards in, 86
  reference substances for, 81—85
  system of, 81—82
  use of, 86
Beverages, see also specific types
  analysis of variance of, 220
Bipolar scales, 208—209, 211, 214
Blank, 152
Blank sheet format, 25—26
Brain hemispheres, 24
Brittleness scale, 95

C

Cake analysis of variance, 221

Canonical Correlation procedure, 36
Categories, number of, 186
Category ratings, 177—190
  analogy to fundamental measurement, 181
  contextual effects, 181—189
  history of, 178—180
  linear relationship between different scales, 186
  practical implications, 189—190
  scales for, 195
    conventional, 199
    examples of, 196
Character notes, 28
Check list, 25
Cheesecakes, texture profiling of, 103—107
Chemicals, as reference standards, 28
Chemical sensory pathways, 23, 32
Chemoreception mechanisms, 147
Chewiness scale, 95
Chi-square evaluation model, 129—142
  statistical significance of taste panel test results with, 131—140
Cleansing palate, 13—14
Cohesiveness of mass, 93—94
Committed responses in writing, 8—9
Comparative judgment, law of, 179—180
Comparison testing, 119—121
Computer analysis, 58—61
Consumer language study, 14—16, see also Language
Consumer panels, 122, see also Panels
Consumers
  preference testing of and sensory evaluation, 166
  sensory registrations of, 15
  sensory testing of, 149—150
Creaminess in foods, 101—103
Creative ability, 12

D

d', 154—157
Data normalization, 204—205
Descriptors, see also Language; Terminology
  for beer, 75—80
  for fat-containing foods, 100
  for fats/oils, 100
  for liquid and semisolid references, 102
Detection threshold, 152
Diaries, 11
Difference
  sensitivity of magnitude estimation to, 220
  testing, 111—123, 150—152
    comparison tests in, 119—121
    correction of preference data following, 140—141
    identification tests for, 112—115
    multiple, 153
    panelists for, 121—122
    possibilities for, 167
    signal detection in, 115—119
Direct scaling, 180

Dual-moderated groups, 6
Dual standard test, 112—113
Duncan Multiple Range Test of Significance, 61
Duo/trio test, 112
  balanced replication and, 50—51
  recordkeeping in, 116
  sensitivity of, 113

E

EBC Thesaurus for brewing industry, 81
Environment control, 13
European Brewery Convention (EBC), 74
Experimental psychologists, 194
Experimenter expectancy effect, 25—26
Expert panels, 122
External modulus normalization, 242
External verbal standards, 212—213

F

Fallis-Lasagna-Tetreault test, 151
Fat-containing foods, texture profiling of, 100—101
Fatigue, 14
  sensory, 82
  taste, 148
Fat-likeness, 99
Fats, 99—101
Fechner's Logarithmic Law, 179
Fechner's scaling methods, 196
Fixed-scale normalization, 242
Flavor, see also Flavor Profile; Taste
  of beer
    reference substances, 83—85
    terminology, 74—86
  definition of, 27
  intensities of, 28—29, 213
  libraries, 81
  variation extremes of, 33
Flavor-by-mouth, 27
Flavorings, 213
Flavor note index, 31
Flavor Profile, 22—23, 28—32
  applications of, 31—32
  conversion of to Profile Attributes, 34
  definition of terms for, 28—29
  establishing final, 29—30
  example of, 30
  final, 29
  format of, 30
  industry use of, 48
  intensity of, 34
  reporting data of, 29—31
  statistical treatment of data of, 31
  uses of, 31
Flavor Wheel, 81, 82
Focus groups, 5
Food colors, 10—11
Food preferences, 149—150
Food vocabulary, 102, see also Descriptors
Force choice, procedures of, 150—151
Forced choice, 126, 152

Fractionation methods, 196
Fragrance substantivity, 226
Frames of reference, 13
Free association, 10
Frequency effects, 182—183
Friedman Analysis of Variance by Ranks, 243
Frozen desserts, evaluation techniques for, 94
Full solution set exercise, 235
Fundamental measurement, 181

G

General Foods Texture Profile Method, see Texture
    Profile Method
Gestalt psychology, 24
Group dynamics, 6—7
  in focus groups, 5
Group Embedded Figures Test, 38—39
Groups, see also Focus groups; Panels
  selecting respondents for, 12—13
  standardization of study design across, 13
  techniques for enhancing productivity of, 8—11
Guessers, elimination of, 127
Guessing, 126
  correction for, 129
Gumminess scale, 95
Gums, texture characteristics of, 101

H

Hardness scale, 95
Harris-Kalmus test, 151
Hedonic curves, 209
Hedonics scale training, 233—234
Home (product) usage, 11
Human senses, 146

I

Identification test, 38
  data analysis in, 114—115
  errors in interpretation of, 114
  test design for, 113—114
  types of, 112—113
Impurities, removal of, 83
In-depth interviews, 5—6
Individual variability, 82—83, 211
Integrative perceptual attributes, 33
Integrative perceptual measures, 32
Intensity, see also Intensity scale; Sensory intensity;
      specific senses
  of different stimuli generating same response,
      205—207
  in flavor profile, 28—29
  measurement of, 240
  monotonic increase in, 199
  rank ordering of, 61
  sensory magnitude and, 199—201
  total, 33
Intensity scale, 28—29, 162, 196—197
  for sweetness, 195
  for texture profiling, 95—96

training, 231—233
Internal standard normalization, 240—241
International Flavors and Fragrances (IFF) Sensory
    Testing Center, 226
    odor discrimination test of, 231
Interval scale, 197
Interviews, in-depth, 5—6
Inverted U curve, 207—208

J

Judges, see also Panelists; Respondents
    computerized tracking of results of, 69
    number of, 57
    performance of, 62—63
    recruitment of, 49
    selection of, 51
    sensory training of, 148—149
Judgmental functions, 182, 188
Judgments
    distribution of, 188—189
    situational dependency of, 181—190
Just noticeable difference (JND), 178—180

L

Language, see also Descriptors; Terminology
    development of, 3—19
        application of to quantitative testing, 17
        form for, 16, 18—19
        group dynamics in, 6—7
        group productivity enhancement techniques for,
            8—11
        instruction, 16
        moderator characteristics for, 7—8
        organization of consumer language study for,
            14—16
        procedure for, 15—16
        in product development, 4—5
        qualitative research designs for, 5—6
        respondent selection for, 12—13
        systematic study design considerations for, 13—
            14
    nonverbal, 11
    respondents' command of, 12
    screening of, 17
    standardized, 150
Leader, see also Moderator
    recommendations and certification of, 40
    styles of, 25
    training and characteristics of, 70
Learning curve, 113—114
Lifestyle (of respondents), 12
Liking
    bipolar rating of, 211, 214
    magnitude estimation of, 207—211
    model for, 207—208
    relation of to sensory attributes, 217
    scales of, 196
        unipolar vs. bipolar, 208—209
Loudness studies, 207
Low-sodium diets, 189

M

Magnitude estimates, 194—222
    advantages and disadvantages of, 227
    averaging of, 213—214
    benefits of, 219—221
    data matrix for ratings of, 206—207
    early studies in, 195—196
    emergence of, 211
    history of, 194—200
    instructions for calibration of, 213
    of liking, 207—211
    limits of, 221
    methodological studies using, 198
    normalizing of, 212—213
        methods of, 239—242
        procedures for, 204
    perceived odor intensity of propanol experiment
        with, 201—207
    power law and, 199—203
    in product testing, 209, 211—215
        sample of, 215—219
    scientific studies using, 197
    standing panels for, 225—248
        data collection and analysis in, 235—243
        judging panelist performance in, 243—247
        panelist motivation in, 247—248
        panelist recruitment and selection of, 227—233
        panelist training for, 231—235
    stimuli-perceptions functional relations and, 199
Malodor masking, 226
Mann and Whitney's U-Test, 163
Margarine evaluation study
    attributes importance measurement in, 217—219
    panel composition and activities of, 215—216
    ratings analysis in, 216
    ratings of self-designed ideal product in, 216—217
    sensory attribute to liking relations, 217
    stimuli for, 215
Marketing, QDA for, 47
Mark sense cards, 235—236
Master Brewers Association of the Americas
    (MBAA), 74
Maxi-groups, 6
Memory drift, 36
Metropolitan Water District of Southern California,
    32
Mini-groups, 6
Mixedness measures, 160
Moderator, see also Leader
    characteristics of, 7—8
    in focus groups, 5
Modulus equalization, 204
Modulus normalization, 204
    external, 242
Mouthfeel, 32
    of fat-containing foods, 99—100
Multiple comparison attribute tests, 121
Multiple comparison hedonic tests, 120
Multiple difference testing, 153
Multiple tastings, 14
Multiplication methods, 196

Multivariate Analysis, 68
Myers-Briggs Type Indicator, 39

N

No difference result, 153
Noise, 153—155
Nonnumerical measurement, 219—220
Nonusers (of product), 12
Nonverbal language, 11
No preference response, 153
Normalization factor, 241
  panelist, 245—247
Normalization methods, limitations of, 242—243
No-standard normalization, 241—242

O

Odor, see also Aroma
  determining intensity and blending of, 226
  divergence of exponents in perception of, 205
  intensity experiment of
    data normalization in, 204—205
    instructions in, 202
    other experimental designs and analysis and,
        205—207
    scaling orientation for, 202—203
    stimuli presentation in, 203—204
    stimuli selection in, 201—202
  recognition series, 38
  weakness, 205
Odor Profiling Method, 48
Off-flavors, 148
Oils, 99—101, see also Fats
Olfactory acuity screening test, 228—231
Olfactory discrimination triangle tests, 229
Open-ended questioning, 8
Order of appearance, 29
Order of perception, 29
Order of presentation, 29
  bias and, 120
  in paired comparison attribute tests, 120
  in paired comparison hedonic tests, 119
  rotated, 118
Orientation, 26
  for magnitude estimation method, 221
  presession, 11

P

P(A), 155—156
  R-index and, 160
Paired comparison attribute tests, 120—121
Paired comparison hedonic tests, 119—120
Paired preference tests, 120
Palate cleansing, 13—14
Panelist normalization factors, 245—247
Panelist(s), see also Judges; Respondents
  motivation of, 247—248
  ranking, 243—245
  scales, 245
Panels, see also Respondents

attrition of, 71
in continuous use, 68—69
feedback for, 70
formal sessions of, 26
intermittently used, 69—70
leaders of, 39—40
  selection of, 25
  training and characteristics of, 70
for magnitude estimation, 225—248
  data collection and analysis in, 235—243
  judging panelist performance in, 243—247
  panelist motivation in, 247—248
  recruitment and selection of, 227—233
  training of, 231—235
maintaining of, 68—70
motivation of, 122, 247—248
obtaining good participation in, 55—57
operating conditions of, 26—28
orientation of, 11, 26, 221
performance of, 63
recruitment of, 49, 96, 227—231
  source of, 122
reference standards of, 27—28
remedial sessions of, 56—57
screening of, 49—52, 97, 122
  tests for, 228—231
selection of, 97, 125—142
  chi-square evaluation model, 129—142
  correction of preference data following a
      triangular test, 129—140
  external, 126—127
  general correction for "guessers", 128
  general data correction following a difference
      test, 140—141
  internal, 127—128
  trainees for, 38—39
standardized smelling and tasting techniques of, 27
time spent in development of, 97
training of, 25, 51, 53—57, 122, see also Training
  programs for, 37—40
  for texture profiling, 96—98
  types of, 122
Parducci's range-frequency model, 188
Pepsi Challenge, 119
Perception(s)
  concept of, 24
  development of scales of, 194
  laws of, 194
    investigation of, 197
  mechanisms of, 194
  order of, see Order of perception
  relations of to stimuli, 199
Perceptual stimuli ratios, 203
Perceptual style tests, 38—39
Physical measurement, rules of, 189
Physical measure-sensory perception relation, 220—
    221
Physical stimuli ratios, 203
Pleasantness ratings, mean, 189
Polygonal tests, 127—128
Power function, 202—203
Power law, 199—203

Power of suggestion, 25—26
Preference testing, 146, 152—153
  sensory evaluation and, 166
Preinterview home usage, 11
Presensitization, 9
Presession orientation, 11
Principal Components Analysis, 36, 99, 103
Probes, 9
Product
  acceptance tests for, 66—67
  descriptions of, replicated, 57—58
  development, common language for, 4
  differences in, 61—62
  home usage of, preinterview, 11
  imagery of, 10—11
  number of, 57
  presentation and serving of, 58
  quantity of, 48
  testing of
    averaging magnitude estimates in, 213—214
    background of, 209—212
    for differences, 112—123
    normalizing magnitude estimates in, 212—213
    sample, 215—219
  usage, 12
    context of, 14
Product evaluation questionnaire, 217
Product imagery, 10—11
Product-language study dynamics, 14—15
Product usage of respondents, 12
Professional respondents, 12—13
Profile Attribute Analysis (PAA), 23, 32—37
  advantages of, 37
  attribute scales for, 34—36
  flavor response sheet for, 35
  panel operation procedures for, 32—33
  selection of attributes for, 33—34
  statistical procedures for, 36—37
  uses of, 37
Profile panel, see Panel
Profiling, 21—40, see also specific types
  attribute analysis, 32—37
  basic principles of, 23—26
  dynamics of, 23—28
  flavor, 28—32, see also Flavor Profile
  methods of, 22—23
    for flavor profiling, 22—23, 28—32
    for profile attribute analysis, 32—37
  panel operation procedures for, 26—28
  panel training programs for, 37—40
  psychophysical issues of, 24
  social psychological issues of, 24—25
  suggestion issues of, 25—26
  texture, see Texture Profile Method
  training issues of, 25
Projective techniques, 10—11
Propanol odor intensity experiment, 201—207
Protomonadic test, 119
Psychographic/lifestyle groups, 12
Psychological measurement scales, 181, 189—190
Psychometric tests, 24
Psychophysical law, 178—179

Psychophysical tests, 24
Psychophysics, 150, 211
  founder of, 194
PTC thresholds, 151

Q

Quadrant rating technique, 55
Qualitative methods for language development, see
  Language, development of
Qualitative research designs, 5—6
Quality control
  informal signal detection testing in, 119
  QDA for, 47
Quantitative Descriptive Analysis (QDA), 43—71
  development of, 44—47
  interpretation of results from, 61—65
  maintenance of trained panels for, 68—70
  maximum effectiveness of, 63—66
  panel leader for, 70
  planning of, 48
  recruitment and panel screening for, 49—52
  replicated product descriptions in, 57—58
  scale for, 44
  scope of industry use of, 47—48
  scoresheet for, 45
  statistical analysis of descriptive data from, 58—61
  strengths and problems with, 70—71
  training sessions and scoresheet development for,
    51, 53—57
  use of with more complex statistical techniques,
    67—68
  uses of data from, 47
    with other tests, 66—67
Quantitative testing, applying language to, 17
Questions
  direct, 11
  exploratory, 14—15
  open-ended, 8
  structuring of for simplicity, 8

R

Randomization, 57
Range-frequency model, 183—185, 188
Ranking test, 38
Rank order technique, 55—56
Rank Sums Test, 163—165
  in R-index value calculation, 167
Rating scales, for texture profiling, 95—96
Ratio scales, 195, 196, 222
  development of, 199
  methodological studies of, 198
Receiver Operating Characteristic curve, see ROC
  curve
Rechner's Psychophysical Law, 178—179
Recognition threshold, 152
Reference standards, 27—28
  for beer, 86
  obtaining of, 35
Reference substances
  for beer, 83—85

for fat-containing foods, 101
for fats/oils, 100
in training of tasters, 86
Regression equation constants, 140
Relaxation training, 10
Replicated pilot test, 56
Replicated product descriptions, 57—58
Replications, number of, 57
Research
    QDA for, 47
    timing of, 13
Respondent-plus-client groups, 6
Respondents, see also Judges; Panelists
    certification of, 40
    chi-square evaluation model for, 130
    diaries of, 11
    for difference testing, 121—122
    general correction for "guessers", 129
    motivation of, 247—248
    performance of, 243—247
    presensitization of, 9
    professional, 12—13
    screening of, 97
        tests for, 38—39, 228—231
    selection of, 12—13, 38—39, 97, 126
        external, 126—127
        internal, 127—129
    sensory registrations of, 15
    time spent in development of, 97
    training of, 231—235
        programs for, 37—40
Responses, enhancement of, 9
Response-Surface Methodology, 36, 68
R-index, 156—162, 167
    application of to consumer preference testing, 166
    computation of, 157—162
        computer program for, 167—173
    intensity scaling and, 162
    mean values of, 166
    rating procedures
        for more than two food treatments, 160—162
        for two food treatments, 157—160
    statistical treatment of values of, 162—166
    as threshold measure, 160
ROC curves, 155—156, 160
Ross Laboratories, 31

S

Saltiness
    mean judgments of, 184
    ratings, 182
Scales, 181, 195
    quality of, 219
Scaling methods, see also specific methods
    development of, 196—197
    early studies of, 198
    importance of, 194
Scoresheets
    development of, 55
    integrating terms into, 56
    transferring of data from to computer, 58—59

Screening tests, 38—39, 48
    conducting of, 51
    setting up, 49—51
Secondary identification, 129
Sensation patterns, 24
Sensory characteristics, 23
Sensory continua, 199—200, 219
Sensory Data System (SDS), 236
    data analysis in, 238—239
    data card entry in, 238
    data report sample for, 240
    entry of ID information in, 236—238
    report generation in, 239
Sensory difference testing, 145—173
    d' in, 156—157
    forced choice in, 150—151
    possibilities for, 167
    testing procedures in, 151—152
Sensory evaluation
    applications of, 146—147
    consumer preference testing and, 166
    d' in, 156—157
    goals and strategies of, 147—150
    magnitude estimation and, see Magnitude
        estimation
    preference and threshold testing and, 152—153
    R-index, see R-index
    signal detection theory, 153—156
    statistical analysis and design of, 165—166
    tests, 150—152
Sensory Evaluation I, 147—148, 167
    R-index values in, 165
Sensory Evaluation II, 148—149, 167
    R-index values in, 165—166
Sensory fatigue, 82, 148
Sensory intensity, 199—200
Sensory measurement
    category scales in, 178
        contextual influences of, 181—189
        history of, 178—180
        practical applications of, 189—190
    psychophysical aspects of, 24
    social psychological aspects of, 24—25
Sensory overload, 14
Sensory perception-physical measure relation, 220—
    221
Sensory psychophysics, 150
Sensory registrations, 15
Sensory research, acceleration of, 199
Sequential effects, 187—188
Show and tell, 11
Signal(s), 153—154
    detection testing of, 115—119, 149
    detection theory of, 115, 146
        ROC curves in, 155—156
        signals and noise in, 153—154
        signal strength measure in, 154
    strength of, 154
Singularity measures, 160
Smell, 32
    standardized techniques of, 27
Smoothness, 103—104

Social psychological issues in profiling, 25
Soft drink taste tests, 113
Sound stimuli, 207
Spider web graph, 44—47, 63—66
Standardization, 13
Standard rating scales, 95
   for teaching and demonstration, 96
Standing panels in magnitude estimation, see
      Magnitude estimation, standing panels for
Starches, texture characteristics of, 101
Statistical design, 57
Statistical software programs, 36—37
Stimuli
   acceptability ratings of, 188—189
   intensity of and sensory magnitude, 199—201
   number of, 187
   paired, 198
   presentation of, 203—204
   range of, 181—182
   ratio evaluations of, 198
   relations with perceptions, 199
   selection of, 201—202
   skewed distribution of, 187
   standard, 198
   transformation of into sense magnitude, 200
Stimulus dimension presensitization, 9
Study design, 13—14
Suggestion, power of, 25—26
Sureness ratings, 155
Sweetness
   perception of, 205—207, 210
   ratings, mean, 189

T

Tactile measurements, 32
Tactile perception, 207
Task orientation, 8
Taste, 32
   basic, 27—28
   divergence of exponents in, 205—207
   interference with, 148
   molecules, reactions of, 147
   preferences, development of new, 189
   receptors for, 147
   vague descriptors of, 4
Taste fatigue, 148
Tasters, training of, 86
Taste tests
   basic, 38
   cleansing of palate in, 13—14
   multiple tasting limits, 14
   statistical significance of, 131—140
Tasting, standardized techniques of, 27
Terminology, 98, see also Descriptors; Language
   changes in, 93—94
Test scheduling, 119
Test sensitivity, 113—114
Tetrade test, 152
Tetrahedral tests, 129
Texture Profile Method, 89—106
   application of to fat-containing foods, 99—106

data analysis of, 98—99
development and applications of, 90—94
evaluation techniques for, 94
evolution of, 93
history of, 91—93
methodology of, 91
nonfood application of, 92
panel training for, 96—98
rating scales in, 95—96
samples of, 93, 104—107
techniques of, 92—93
terminology changes in, 93—94
vocabulary for, 98
Thirteen-sampler odor set exercise, 234—235
Three-alternative forced choice (3-AFC) test, 151,
   156
Threshold measure, 160
Threshold testing, 146, 152—153
Thurnstonian scaling, 179—180
Thurstone's Law of Comparative Judgment, 179—
   180
Time and responsibility schedule, 48
Training
   for magnitude estimate panel, 231—235
   of panel leader, 70
   planning of, 53—54
   products needed for, 53
   in profiling, 25
   sample planning sheet for, 54
   scoresheet development and, 55
   for texture profiling, 96—98
   time schedule for, 53
Training products, 48
   types of, 53
Triangle difference test results, 52
Triangle test, 50, 112
   correction of preference data following, 130—140
   data analysis of, 115
   in determining respondent sensitivity, 127
   false alarm rate for, 116
   for panelist screening, 229—230
   sensitivity of, 113
Trihedral tests, 129

U

Unipolar scales, 208—209

V

Verbal fluency, 12
Verbal standards, external, 212—213
Viscosity, 94
   scale for, 95
Visual measurements, 32

W

Warm-up questions, 7
Weber's law, 178
Wilcoxon Rank Sums Test, 163
Written responses, 8—9